T0214435

Environmental Engineering: Review for the Professional Engineering Examination

Ashok Naimpally · Kirsten Sinclair Rosselot

Environmental Engineering: Review for the Professional Engineering Examination

 Springer

Ashok Naimpally
Division of Math Science and Engineering
Fresno City College
Fresno, CA
USA

Kirsten Sinclair Rosselot
Process Profiles
Calabasas, CA
USA

ISBN 978-1-4899-7894-3   ISBN 978-0-387-49930-7   (eBook)
DOI 10.1007/978-0-387-49930-7
Springer New York Heidelberg Dordrecht London

Printed on acid-free paper

(Corrected at 2nd printing 2013)

Springer is part of Springer Science+Business Media (www.springer.com)

Preface

This book provides a review of key concepts and theory as well as solved problems for candidates seeking to prepare for the Principles and Practice of Engineering Examination in the area of Environmental Engineering. Passing this exam is one of the requirements for obtaining a professional license for environmental engineering, or PE Environmental. The examination is prepared by the National Council of Examiners for Engineering and Surveying (NCEES), and the examination syllabus and other details are described on the NCEES website.

The best way to study for an engineering exam is to practice solving problems. In this book, two practice exams with 100 problems each are presented, and solutions to each problem are provided so that the candidate can check their answers. A brief review of each of the environmental engineering topic areas covered in the exam is given in order to refresh the candidate's memory of concepts they might not have encountered since they left school. While this book does not attempt to serve as a substitute for standard textbooks and reference books in environmental engineering, it could serve as a useful adjunct to such texts.

Ashok Naimpally wishes to acknowledge his wife Shobha, parents Radha and Vasudev, and family and friends. Kirsten Sinclair Rosselot also wishes to acknowledge family and friends who supported and encouraged her during the preparation of this book. They would also like to thank the staff at Springer for their patience and editorial assistance.

August 9, 2013

Contents

Chapter 1
Water and Wastewater

1.1 Water Resources Planning

1.1.1 Population Projection

Planning for water and other material and energy resources is dependent upon two main factors:

1. Population.
2. Per capita consumption of the commodity (water, energy, and resource) or per capita depletion of environmental resources.

The population trends for the past two centuries can be summarized as follows [1]:

1. Urbanization or the trend to move from rural to urban areas.
2. Industrialization or the rapid means of manufacture using technological innovations.

Both urbanization and industrialization are interrelated though not the same.

The four factors that affect the population growth rate, r, in a given geographical area are the following: birth rate b; death rate d; immigration rate i; and emigration rate m. The growth rate r can be given as a function of these factors by the equation [2]:

$$r = b - d + i - m$$

Both mathematical methods and methods of graphical projection are used in coming up with population forecasts, which is a complex task in its own right [3].

The population projections can be done using two main types of models:

1. Geometric projection. This is expressed as an exponential growth equation

$$P_t = P_o \exp(rt) \tag{1.1}$$

A. Naimpally and K. S. Rosselot, *Environmental Engineering: Review for the Professional Engineering Examination*, DOI: 10.1007/978-0-387-49930-7_1,
© Springer Science+Business Media New York 2013

P_t Population at time t
P_o Population at time $= 0$
r   a constant

2. Linear projection.

The population is expressed as a linear equation with time:

$$P_t = P_o + At \qquad (1.2a)$$

A   a constant

Or, a composite model

$$P_t = P_o + At + P_o \exp{(rt)} \qquad (1.2b)$$

Of the two methods, the geometric growth method fits into a natural scheme of things. More complicated models like (1.2b) can be useful.

A Population Pyramid shows the percent population as a function of age. This can be done for both males and females. Fertility rate is the number of children (average) born per woman. Zero growth occurs when the death rate equals birth rate, and there are no immigration or emigration. Per capita consumption for different age groups can also be taken into account in planning for resources.

1.1.2 Legal Aspects of Water Rights [4, 5]

Water rights are often the cause of conflicts within states, countries, and nations. A summary of the different kinds of water rights is given below:

Riparian rights refer to the rights of persons whose land is adjoining a body of water.

The Appropriation doctrine refers to the "first in time, first in line" doctrine developed during the gold rush period in California.

Groundwater rights are more complicated. The English rule states that an owner of land has absolute jurisdiction over the groundwater below. The American principle is one of reasonable use and that the rights of the landowner are limited.

Federal rights in the U.S. Constitution, although limited to navigation in Article 1, actually has been interpreted broadly due to case law, and the federal government has a lot of leeway over America's water. Federal Reserved Water Rights refer to the water needs of National Parks, National Forests, etc.

Tribal Water Rights refer to rights of Native Americans arising from treaties and water adjoining tribal lands.

A large number of federal and state statutes need to be considered in water resource planning, the main one being the Safe Water Drinking Act.

The Clean Water Act (1971) makes use of a variety of regulatory and non-regulatory methods to reduce pollutants that are discharged, finance municipal wastewater treatment plants and manage polluted runoff. The Act's stated objective is to "restore and maintain the chemical, physical, and biological integrity of the nation's waters." The Act achieves these objectives by establishing the following goals:

1. Elimination of the discharge of pollutants into national waters.
2. Achievement of a level of water quality, which provides for the protection and propagation of fish, shell fish, and wildlife and for recreation in and on water.

As authorized by the Clean Water Act, the National Pollutant Discharge Elimination System (NPDES) permit program controls water pollution by regulating point sources that discharge pollutants into waters. In most cases, the NPDES system is administered by the states. Point sources are discrete conveyances such as pipes or man-made ditches. Industrial, municipal, and other facilities must obtain permits if their discharges go to surface water.

Part of the Clean Water Act focuses on chemical aspects and point source facilities. The EPA has developed over twenty industrial categories based on wastewater effluent quality. For each of the industrial categories, the EPA has set forth specific regulations and effluent limitations.

Safe Drinking Water Act (SDWA):
The perspectives that have been incorporated in the SDWA are:

1. Water is internally consumed and required for life.
2. Water utilities may be more expensive than other utilities, since the drinking water standards need to be maintained.

The original Act had two basic thrusts:

1. Required the EPA to establish national standards for drinking water.
2. Required operations of water utilities to continuously monitor water quality.

Two sets of water standards are set:
National Primary Drinking Water Regulations (NPDWR) set the maximum contaminant levels (MCL) in water. The MCLs are set by the EPA while taking into account the best available technologies and measuring instrumentation. The NPDWR are based on the need to protect human health and EPA uses this criterion.

- National Secondary Drinking Water Regulations (NSDWRs). These are voluntary standards and were based upon esthetics like odor, taste, and color.

The 1996 amendments to SDWA provided for:

- Identification, monitoring, and control of water contaminants.
- Enforcement of rules.
- Utility collection of information and dissemination of information.
- Provided so called customer "right to know."
- Funding for utilities to upgrade plants.

1.1.3 Sources of Water

1. Rivers, lakes: Conflicts arise from construction of dams with rights of upstream versus downstream users. Water purification plants need to be built for water being supplied to cities, towns, villages, etc.
2. Ground water is an important source. Appropriate purification steps are needed.
3. Sea: Desalination is an emerging large-scale technology, especially with the newer reverse osmosis process being far more economical than the older distillation and freezing processes.
4. Use of icebergs has been explored from time to time.
5. Reclaimed water obtained from wastewater treatment plants is being gradually accepted for uses such as watering of lawns and plants.

1.1.4 Uses of Water

1. Human (domestic).
2. Industrial.
3. Agricultural.
4. Recreational.
5. Power generation.
6. Other.

Agricultural usage is considered consumptive, since a large percent of water (up to 85 %) is lost to the atmosphere due to evaporation. On the other hand, water for industrial uses (and to a limited extent, domestic uses) can be used for other uses after appropriate treatment.

1.1.5 Large-Scale Planning

Large-scale planning (for a city or a region) involves several sources of water, several end-uses as well as formulating goals, evaluation of alternatives using known data and mathematical models. There are many uncertainties both in the qualitative factors and the mathematical models used for the technological and population aspects of the planning process.

1.1.6 Other Factors in Planning

The ecological effects of water use, such as loss of wildlife due to construction of dams, and the political factors coupled with intense emotions on the subject can lead to conflicts.

1.2 Water Distribution and Wastewater Collection

1.2.1 Intake of Water for Distribution

From reservoirs:

Water is obtained from reservoirs by means of intake pipes located below the water level of the reservoir, but well above the floor to avoid problems of silt finding its way into the water supply.

From rivers:

The intake of water supply from a river can be either by means of inlet pipes drawn from a weir located in the river, or an intake well in the middle of the river, or intake with the help of a canal from the river with the intake well located in it.

From underground wells:

Wells are drilled either in sand beds or in ground beds. The outer casing of the well, which is set in place after drilling, is sealed with concrete being poured between the casing (made up of iron or steel or PVC) and the soil. A screen is used to obtain the water but keep out the sand or gravel. A pump at ground level pulls up the water and discharges it into the piping system.

1.2.1.1 Transportation of Water

Transportation of water from the source to the population centers can be made to occur by two methods, gravity flow, and flow due to pressure drop (pumping).

1. Gravity flow

Gravity flow utilizes the hydraulic gradient and thus avoids sharp changes in contours like hills and mountains. The flow channel may thus have to take a longer route.

Gravity flow can be made to occur in the following:

- Canal, either lined or unlined.
- Aqueduct, which is an artificial channel.

Tunnels may need to be drilled at an appropriate place in case of mountains, hills, etc. Since the surface of the water is open to the atmosphere, loss of water due to evaporation could occur, as also pollution from surface water runoff.

2. Flow due to pressure drop (pumping).

A pressurized flow system has the advantages of being on the direct route (even if the route can have an uneven hydraulic gradient) and thus costing less for upkeep.

Materials of construction for the piping system

The possible materials of construction are ductile iron, steel, concrete, or plastics.

Iron has the advantages of strength, optimum cost, and low corrosion rates. However, large pressures may be fatal to the pipe.

Steel pipes are strong and have low weight. However, they have the disadvantage of not being corrosion resistant to variations in the pH of water, and of not withstanding negative pressure due to vacuum in the pipe. The life of steel pipe is much less than that of iron.

Concrete pipes are not biodegradable, can be constructed on the site, and can resist external compressive loads. They have the disadvantages of not being resistant to acids or alkalis and of lack of ease of repairs.

Plastic pipes, generally made of polyethylene or PVC, have the advantages of being lightweight, having corrosion resistance, and having good strength. Some types of plastic pipes made of PVC are not suitable for hot water systems.

Piping systems within a building are made up of copper, plastic, or brass.

1.2.1.2 Sea Water

Seawater is also used as a source of water. The water can be desalinated with any of these processes.

1. *Distillation.* Costly, not popular.

   The next 2 processes, electrodialysis and reverse osmosis, are membrane processes. A membrane is a thin microporous sheet controlling the passage of molecules based on shape, size. or charge of molecule [6].

   They depend upon the ability of membranes to allow (or conversely, disallow) certain types of ions to move across the membrane. A good fundamental treatment of diffusion across membranes and of membrane based processes is given by Cussler [7].

2. *Electrodialysis.*

   Electrodialysis is a process involving a membrane technique for: separating salts, acids, and bases from aqueous solutions; the separation of monovalent ions from multivalent ions; and the separation of ionic compounds from uncharged molecules [8]. The ion-solution membrane acts as a physical barrier through which ions are transported away from a feed solution. Electrodialysis is used to obtain potable water from brackish water, nitrate removal from drinking water, etc.

   The electrodialysis membrane stack consists of electrodes and membranes separated by gaskets and spacers in a manner that promotes turbulence in the compartments in order to reduce the resistance to the flow of electricity.

In the electrochemical process, direct (DC) current is made to pass through the process. The positively charged ions (cations) move toward the cathode (e.g., Na^+); and the negatively charged ions (anions) move toward the anode (e.g., Cl^-). A large number of anodes and cathodes are placed alternately in the electrodialysis cell. The cations can pass through the cation-permeable membrane but are obstructed by the anion-permeable membrane. Vice versa for anions.

The current needed can be computed from the formula [9]

$$I = \left(\frac{FQN}{n}\right)\left(\frac{E_1}{E_2}\right) \tag{1.3}$$

where
I Current in amps
Q Flow rate in liters/sec
N Number of cells
n Normality of the solution in g.moles/L
E_1 Removal efficiency (fraction)
E_2 Current efficiency (fraction)

If the total resistance of the cell is R in ohms, the voltage E is

$$E = IR. \tag{1.4}$$

Power in watts consumed by the electrodialysis cell is

$$P = I^2R. \tag{1.5}$$

3. *Reverse osmosis* (Fig. 1.1).
   Osmosis is a natural phenomenon of establishing thermodynamic equilibrium between 2 solutions; in the case of two solutions separated by a membrane, the osmotic pressure is the extra pressure that must be applied on the solution in order to check the inflow of solvent and establish thermodynamic equilibrium.

Fig. 1.1 A continuous reverse osmosis unit

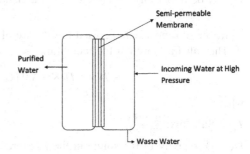

Semi-permeable Membrane

Purified Water

Incoming Water at High Pressure

Waste Water

Reverse osmosis is an operation in which pressure higher than the osmotic pressure is applied so that solvent diffuses from the concentrated solution through the membrane into the dilute solution [6]. The thermodynamic efficiency of the process is high, since the energy expended for the process in the form of applying pressure can be recovered.

A semi-permeable membrane (like cellulose acetate or polyamide polymer) when used to separate a concentrated salt solution from a dilute one causes a flow of solvent from dilute solution to the concentrated one. The osmotic pressure that develops is given by the equation [9].

$$\pi = \phi v \frac{n}{V} RT \tag{1.6}$$

where

π Osmotic pressure
ϕ Osmotic coefficient
v Number of ions produced from a molecule of the electrolyte
n Number of moles of electrolyte
V Volume of solvent
R Gas constant
T Absolute temperature in K

In the normal course, the solvent flows from the dilute to the concentrated solution. If, however, hydrostatic pressure in excess of the osmotic pressure is applied on the concentrated side, the solvent is pushed from the concentrated side to the dilute solution. In the case of brine solution, the flux of water is given by the formula [9, 10]:

$$J_w = W_p(\Delta P - \Delta \pi) \tag{1.7}$$

J_w Water flux through the membrane g.moles/cm^2 area—second
ΔP Pressure differential across the membrane $= P_{IN} - P_{OUT}$ (atm)
$\Delta \pi$ Osmotic pressure difference across the membrane $= \pi_{in} - \pi_{out}$ (atm)
W_p Coefficient of water permeation, a characteristic of the membrane

Reverse osmosis can be used in the case of both ionic and non-ionic solutions. The salt flux through the membrane is

$$J_S = (D_S K_S / \Delta z)(C_{IN} - C_{OUT}) \tag{1.8}$$

where

J_S Salt flux, $\frac{\text{g.mols}}{\text{cm}^2\text{sec}}$
D_S Diffusivity of the solute in the membrane, cm^2/sec

K_S Solute distribution coefficient (dimensionless)

C Concentration $\frac{\text{g.mols}}{\text{cm}^3}$

Δz Thickness of membrane

The term $\frac{D_S K_S}{\Delta z}$ is traditionally called K_P, the membrane solute mass transfer coefficient.

1.2.2   Water Distribution and Transportation

Enough pressure needs to be provided in the water mains in order to meet the requirements of firefighting, as well as the needs of domestic, industrial, and commercial customers. Residential requirements can vary from hour to hour, day to day, season to season, and from year to year.

The Fire Suppression Rating Schedule published by the Insurance Services Office (ISO) gives the process for designing, constructing, and maintaining water requirements for firefighting.

The grid system of pipe networks is the preferred one for designing pipe systems in a city or town. Appropriate amount of water storage must be provided by using either elevated or ground level storage, or underground storage tanks. Provisions must be made for backflow prevention, to avoid contamination of water supply from unforeseen industrial or residential services. Elevated tank storage is useful for the contingencies arising from cutoff of electrical power. In addition to backflow prevention, automatic valves need to be installed to isolate any section of the piping network that may have a fire in the section.

The amount of storage capacity needed can be computed graphically from the hourly water consumption data. The procedure for computing the minimum amount of storage capacity involves the following steps [11]:

1. Obtain the data for water usage as a function of time, preferably on the day of maximum water usage in the city.
2. Convert the raw data into the cumulative amount of water used up until time t.
3. Plot a graph of the data with time t on the X-axis and entire cumulative water usage till time t. A sample plot is shown in Fig. 1.2.
4. The point of maximum usage (at $t = 24$ h) is given by point M on the graph.
5. Draw a straight line going the origin 0 to the point M. This shows the hypothetical flow indicating the cumulative usage of water in the event of the constant rate of usage throughout the day, with the constant rate being equal to the average rate of water usage.
6. Draw two lines parallel to the straight line OM, as distinct as possible, both above and below that line in the following manner:
(a) At the point of inflexion M_{up}, the tangent is a line parallel to OM and as far away from OM as possible and above OM. The line is called line 1.

Fig. 1.2 Flow mass curve
(Rippl diagram)

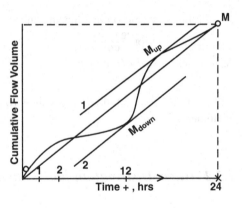

(b) At the point of inflexion M_{down}, the tangent is a line parallel to OM and as far away from OM as possible and below OM. The line is called line 2.

7. Obtain intercepts of lines 1 and 2 on the Y-axis at I_1 and I_2. The minimum amount of storage is equal to the value $(I_1 - I_2)$.

1.2.2.1 Design of Hydraulic System

The design of the flow conduits for water is based on the following formulas used for open channel flow and/or for pipe flow. The parameters for flow conduits are allowable velocity, available pressure or head, and the given resistance to flow. By convention, the Manning's equation is used for open channels and the Hazen–Williams equation is used for pressure conduits [12].

1. Manning's equation

$$V = \frac{k}{n} R^{2/3} S^{2/3}$$

k 1 for S.E. units
k 1.486 for USCS units
V Velocity of flow (m/s or ft/s)
n Roughness coefficient
R Hydraulic radius (m or ft) (defined later)
S Slope of energy grade line (m/m or ft/ft)

2. Hazen–Williams equation

$$V = k_1 \, CR^{0.63} \, S^{0.54}$$

C Roughness coefficient
k_1 0.849 for S.I. units
k_1 1.318 for USCS units

1.2.2.2 Hydraulic Radius R Used for Non-circular Conduits

For a non-circular closed conduit or an open channel, the hydraulic radius R is defined by

$$R = \text{cross-sectional area of fluid flow/wetted perimeter}$$

1.2.3 Wastewater Collection Systems

Before the design of a wastewater collection system, the amount of sewage flow should be estimated from the following parameters.

1. *Population.* The historical data for amount of sewage versus the population must be analyzed.
2. *Water usage.* About 70–80 % of the water used by the population in the town is expected to come back into the sewage system.
3. *Water infiltration.* Even when the sewage system and the storm water system are left separate, a certain amount of infiltration of water into the sewage system is to be expected.

The amount of sewage in the system can fluctuate hourly, daily, and monthly. For sewers, the design basis is to have the sewer lines run 75 % full for maximum daily flow. The maximum hourly flow is important, since the pressure of solids in the sewers can cause attrition of the sewers; the maximum hourly flow is taken to be thrice the average daily flow. The minimum flow in the sewer is also important, because clogging should not occur at minimum flow. Minimum flow is taken to be 66 % of the average hourly flow.

The data from various sources are then analyzed and put in the form of graphs based on curve-fit formulas, for the minimum and maximum flows as functions of population. The graphs give the ratio of minimum to average flow, as well as the ratio of maximum to average flow.

The traditional hydraulic formulas for estimating flow velocities in sewers are the Chezy's formula, Bazin's formula, Manning's formula, and Hazen–Williams' formula.

Manning's Equation

$$V = (k/n)R^{2/3}S^{1/2} \tag{1.9}$$

where
k 1 for SI units
k 1.486 for USCS units
V Velocity (m/s, ft/sec)
n Roughness coefficient
R Hydraulic radius (m, ft), and
S Slope of energy grade line (m/m, ft/ft)

Hazen–Williams Equation

$$V = k_1 C R^{0.63} S^{0.54} \tag{1.10}$$

where
C Roughness coefficient,
k_1 0.849 for SI units, and
k_1 1.318 for USES units.

For pipes with constant roughness, formulas are available for relating the ratios of [13, 14]: depth d to diameter D of the pipe; velocity when flow occurs at depth d versus full flow at diameter D; area A at d versus A_f at D; flow rate Q at d versus Q_f at D; and hydraulic radius R at d versus R_f at D. Values of d, D, and α are shown in Fig. 1.3

$$\frac{d}{D} = \frac{1}{2}\left(1 - \frac{\cos\alpha}{2}\right) \tag{1.11}$$

$$\frac{v}{V} = \left(1 - \frac{360° \sin\alpha}{2\pi\alpha}\right)^{2/3} \tag{1.12}$$

Fig. 1.3 Partially filled circular sewer section

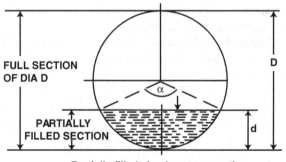

Partially filled circular sewer section.

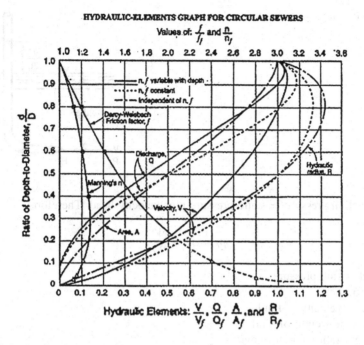

Fig. 1.4 Hydraulic elements graph

$$\frac{A}{A_f} = \left(\frac{\alpha}{360^\circ} - \frac{\sin \alpha}{2\pi}\right) \tag{1.13}$$

$$\frac{Q}{Q_f} = \left(\frac{\alpha}{360^\circ} - \frac{\sin \alpha}{2\pi}\right)\left(1 - \frac{360^\circ \sin \alpha}{2\pi\alpha}\right)^{2/3} \tag{1.14}$$

$$\frac{R}{R_f} = \left(1 - \frac{360^\circ \sin \alpha}{2\pi\alpha}\right) \tag{1.15}$$

These formulas are graphed in the form of plots with d/D on the Y-axis and the other four ratios on the X-axis. The graph is shown in Fig. 1.4.

1.2.4 Fluid Mechanics and Hydraulics

1.2.4.1 Flow Measurement Devices

1. Pitot tube (Fig. 1.5).

The pitot tube has a small tip that senses the stagnation pressure at the tip. The stagnation pressure is the total head of the fluid at the location of the tip.

Fig. 1.5 Pitot tube

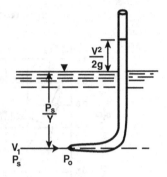

Stagnation pressure = static pressure plus velocity head. Therefore,

$$p_o = p_s + \frac{\gamma V^2}{2g} \tag{1.16}$$

p_o Stagnation pressure
p_s Static head
V Local velocity of fluid
γ Specific weight of fluid
g Acceleration due to gravity

The formula for the venturi meter is:

2. *Venturi meter*: Venturi meter is a device inserted in a pipe that has a short
   length of convergent pipe followed by a large length of divergent pipe. The
   convergent pipe increases the fluid velocity. The divergent section of the
   Venturi meter recovers most of the losses in the convergent section of the pipe.
   This fact is reflected in the coefficient of velocity C_V having the value of 0.98.
   Pressure taps at points 1 and 2 measure the pressures at point 1 and 2.

$$Q = \frac{C_V A_2}{\sqrt{1 - (A_2/A_1)^2}} \sqrt{2g\left(\frac{P_1}{\gamma} + z_1 - \frac{P_2}{\gamma} - z_2\right)}, \tag{1.17}$$

where
C_V The coefficient of velocity = 0.98

The above equation is for *incompressible fluids*.
P_1 Pressure at point 1
P_2 Pressure at point 1
A_1 Area of cross-section at 1
A_2 Area of cross-section at 2

Table 1.1 Orifices and their nominal coefficients

Orifices and their nominal coefficients		
	Sharp edged	Rounded
C	0.61	0.98
C_C	0.62	1.00
C_V	0.98	0.98

3. *Orifice meter*: The orifice meter is a low-cost flow measuring device. An orifice, which can be fabricated in-house, is inserted in the pipe. The permanent pressure loss is greater in the orifice meter than in the Venturi meter—there is no divergent section for energy recovery.

The cross-sectional area at the vena contracta A_2 is characterized by a *coefficient of contraction* C_C and given by $C_C A$ (Table 1.1)

$$Q = CA\sqrt{2g\left(\frac{P_1}{\gamma} + z_1 - \frac{P_2}{\gamma} - z_2\right)} \qquad (1.18)$$

where C, the *coefficient of the meter*, is given by

$$C = \frac{C_V C_C}{\sqrt{1 - C_C^2 (A/A_1)^2}} \qquad (1.19)$$

Values of the coefficients C_V and C_C are given in Table 1.1.

1.2.5 Fluid Flow in Piping Systems

Case 1: Flow in simple, straight pipes (Fig. 1.6)

Consider a section of pipe between the pipes *a–a* and *b–b*.

The final equation for the conservation of energy (otherwise known as the modified Bernoulli's equation) can be written as (for an incompressible fluid):

$$\frac{p_a}{\rho} + \frac{\alpha_a u_a^2}{2} + gZ_a + \eta W_p = \frac{p_b}{\rho} + \frac{\alpha_b u_b^2}{2} + gZ_b + h_f \qquad (1.22)$$

Fig. 1.6 Flow in straight pipes

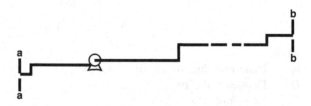

where each of the terms stands for the following:

p_a Pressure of fluid in the pipe at point a
p_b Pressure of fluid in the pipe at point b
u_a "average" velocity of fluid in the pipe at point a
u_b "average" velocity of fluid in the pipe at point b
α_a Kinetic energy correction factor at point a
α_b Kinetic energy correction factor at point b
Z_a Elevation of fluid at point a
Z_b Elevation of fluid at point b
ρ Density of fluid
W_p Electrical energy fed to the pump per unit mass of fluid flowing in the pipe
η Efficiency (overall) of the pump
h_f Energy loss in the pipe due to friction per unit mass of liquid flowing in the pipe

A few elementary definitions are in order:

$$t = \mu \frac{du}{dy} \tag{1.23}$$

where

t Shear stress in the fluid
μ Viscosity of fluid
du/dy Shear rate in the fluid

The Fanning friction factor, f, for flow in a pipe of circular cross-section is defined by:

$$f = \frac{\tau_w}{\rho \overline{V}^2 / 2g} \tag{1.24}$$

where

τ_w Shear stress at the wall
$\overline{V}$ Average velocity of flow in the pipe
ρ Density of fluid

The conservation of momentum equation, applied to fluid flow in a pipe of circular cross-section, yields the following relations:

$$h_{fs} = \frac{4fLV^2}{2gD} \tag{1.25}$$

$$\Delta p_s = -(\rho)(h_{fs}) \tag{1.26}$$

where

h_{fs} Head loss due to skin friction *only*
D Diameter of pipe
L Length of pipe

V Average velocity of fluid in the pipe
ρ Density of fluid
Δp_s Pressure differences due to fluid friction
and,

f = Fanning friction factor
  = for laminar flow, $f = 16/N_{Re}$
    For turbulent flow, f must be found from the friction factor chart

$$N_{Re} = \text{Reynolds number}$$
$$= DV\rho/\mu$$

1.2.5.1 Computation of h_f, the Head Loss

In addition to the formula given above for the computation of head loss due to skin friction, one can use the following formulas for the computation of head loss due to changes in the orientation/size/shape of pipes [6]:

1. Expansion of pipe (Fig. 1.7)

$$h_{fe} = K_e \overline{V}_a^2/2g \tag{1.27}$$

where $K_e = (1 - S_a/S_b)^2$ where S_a = cross-sectional area of pipe at point a and S_b = cross-sectional area of pipe at point b.

2. Contraction of pipe (Fig. 1.8)

$$h_{fc} = K_c \overline{V}_b^2/2g \tag{1.28}$$

Fig. 1.7 Expansion of pipe

Fig. 1.8 Contraction of pipe

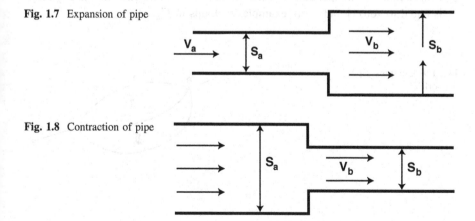

Table 1.2 Table of fitting coefficients

	K_f
Globe valve, wide open	10.0
Angle valve, wide open	5.0
Gate valve, wide open	0.2
Half open	5.6
Return bend	2.2
Tee	1.8
Elbow, 90°	0.9
45°	0.4

where $K_c = 0.1$ for laminar flow and $K_c = 0.4(1 - S_a/S_b)$ for turbulent flow and where S_a = cross-sectional area at point a and S_b = cross-sectional area at point b.

3. Fittings and valves

$$h_{ff} = K_f \overline{V}_a^2 / 2g \tag{1.29}$$

where

K_f Loss factor for fitting
$\overline{V}_a$ Average velocity in pipe leading to fitting

The value of K_f can be found from Table 1.2.

Case 2. Complex piping network

A complex piping network can be divided into several loops (an example is shown in Fig. 1.9).

The two principles upon which the analysis of the complex network is conducted are:

1. Conservation of mass: The total inflow to a node is equal to the total outflow from a node (a, b in Fig.1.9 are examples of nodes).
2. Conservation of momentum: The net pressure drop in moving around any loop is equal to zero (I and II are examples of loops in Fig.1.9).

Fig. 1.9 Complex piping network

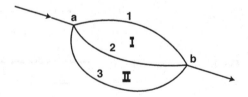

1.3 Characteristics of Water and Wastewater

A classification of the various kinds of pollution found in domestic and industrial wastewaters and in surface runoff is given in the Table 1.3.

1.3.1 Chemical Types of Pollution

Pollution by organic compounds: Most of the organic compounds can be broken down by micro-organisms present in surface water bodies through biochemical reactions utilizing the dissolved oxygen in the water. Even toxic substances can suffer breakdown in this way provided they are present in sufficiently low concentrations. There are, however, a number of organic compounds that are resistant to microbiological decompositions, for example, hydrocarbons, ethers, some vinyl compounds, alkylbenzene sulfonates (ABS), cellulose, and coal.

If the pollution load is small and the dilution by well oxygenated stream water is high, sufficient dissolved oxygen may be present to enable aerobic bacteria to break down the organic matter completely to stable, relatively harmless and odorless end-products. Through this process of stabilization, the water body thus recovers naturally from the effects of pollution and is said to have undergone "self purification." The oxidation reactions which occur are as follows:

$C \rightarrow CO_2$ + carbonates

$H \rightarrow H_2O$

$N \rightarrow NH_3 \rightarrow$ nitrites $\rightarrow$ nitrates (nitrification)

$S \rightarrow$ Sulfates

$P \rightarrow$ Phosphates

Often, however, massive pollution by organic matter causes exhaustion of the dissolved oxygen. The remaining organic matter is then broken down by anaerobic bacteria which do not require free oxygen but can utilize combined oxygen in the form of nitrates, sulfates, phosphates, etc. Putrefaction then occurs, resulting in the breakdown of organic matter to a different set of end products, some of which have objectionable odors. The anaerobic decomposition of organic matter requires

Table 1.3 Types of pollutants

A. Chemical	B. Physical	C. Physiological	D. Biological
Organic	Color	Taste	Bacteria
	Turbidity		Viruses
Inorganic	Suspended and floating matter	Odor	Fungi
			Algae
	Foam		Parasitic
	Temperature		Worms
	Radioactivity		

the participation of two groups of anaerobic bacteria, namely acid- and methane-producing bacteria, resulting in the production of organic acids and methane, respectively; also, new cells are produced. The oxidation and reduction reactions occurring in the anaerobic breakdown or organic matter are as follows:

C → Organic acids (R–COOH) → CH_4 + CO_2
N → Amino acids (R·NH$_2$·COOH) → NH_3 + amines
S → H_2S + organic S-compounds
P → PH_3 (phosphine) + organic P-compounds

Pollution by inorganic chemicals. Many industrial wastes contain corrosive inorganic acids or alkalis that can do extensive damage to aquatic life by breaking down its natural buffer system and altering its normal pH. Acids and alkalis can destroy bacteria and so inhibit self-purification of a stream; they are also lethal to fish and other forms of aquatic life.

Some industrial wastes contain toxic inorganic compounds such as free chlorine, ammonia, hydrogen sulfide, and soluble salts of heavy metals (copper, zinc, chromium, cadmium, mercury, etc.). Any appreciable amounts of these compounds can destroy aquatic life, whether animal or vegetable.

1.3.2 Physical Types of Pollution

Color. Many industrial wastes discharging to a stream have a pronounced color which they may impart to the water. The color is due, in most cases, to organic dyes, but there are some highly colored substances of mineral origin, especially compounds of iron and chromium.

Interaction between an industrial effluent and substances present naturally in a stream can produce intense coloration. One example of this is the reaction between iron-containing mine effluent and the natural bicarbonate alkalinity of a river to give a reddish brown opalescence and eventually a deposit of ferric hydroxide, $Fe(OH)_3$.

Turbidity. A striking physical characteristic of domestic sewage and of most industrial wastewaters is their degree of cloudiness or turbidity, which is caused by the presence of colloidal matter. Colloidal particles are in the size range of 1–1000 nm (1 nm = 10^{-7} cm). They are too small to be retained by ordinary filter paper; they do not settle on standing. Colloidal particles have a very large surface area per unit volume. (A 1 cm cube having a total surface area of 6 cm^2, if subdivided into 8×10^{12} cubes, will have a total surface area of 12 m^2.) This vast surface area is electrically charged, usually negatively, and requires positively charged ions for the neutralization of the surface charge, resulting in coagulation to coarse particles.

Thermal pollution. Temperature as a form of water pollution has both industrial and ecological significance. The discharge of heated process water from industrial plants and turbine cooling water from electric generating plants may easily cause a temperature rise of several degrees in a river or stream. When a rise in temperature occurs in a stream polluted by organic matter, there is not only disappearance of dissolved oxygen due to the lower solubility of oxygen at the higher temperature, but also an increased rate of utilization of dissolved oxygen by biochemical reactions, which proceed much faster at higher temperatures. For these reasons, many streams may be satisfactory concerning dissolved oxygen content in the winter, but may contain little or none during the summer.

If the dissolved oxygen does fall to zero, putrefaction of organic matter will occur, giving rise to bad odors and nuisance.

Although fish can become somewhat acclimatized to high temperatures, above a certain temperature they will eventually die. Lethal temperatures for trout, pike, and goldfish are 25, 30, and 35 °C, respectively.

Suspended solids. Insoluble matter in suspension is one of the commonest forms of water pollution, being present in most industrial wastewaters and in domestic sewage.

Suspended solids are objectionable in streams because (a) they interfere with self-purification by diminishing photosynthesis, (b) they can damage fish life, and (c) they are unsightly.

Foam. Foam consists of a dispersion of gas bubbles (usually air) in a water medium. It became a serious problem with the increasing use of synthetic detergents, of which the alkylbenzene sulfonates are the most widely used for domestic purposes. Unfortunately, these are the most difficult to break down during sewage treatment, and consequently, about 50 % of the amount originally present in raw sewage passes through the treatment plant into the receiving streams, thus producing masses of foam.

The production of foam is a physical phenomenon due to the lowering of the surface tension of water by the detergent.

The development of new, biodegradable detergents have eased this problem.

Radioactivity. Power production by nuclear fission resulting in large volumes of radioactive wastewaters is responsible for most of the water pollution by radioactivity since fallout from nuclear weapon testing in rainfall has been discontinued by the major powers (except by China and France). Sewage treatment has varying effects on different isotopes. Some isotopes, such as Sr^{90}, tend to pass through the treatment plant, whereas others, such as Pu^{239}, are adsorbed in the sludge.

When a radioactive effluent is discharged to a river, the radioactivity of that river will depend on the degree of dilution. In addition, the activity will tend to diminish in proportion with the distance from the source of pollution because of (1) sedimentation, (2) decay of the short-lived isotopes, and (3) uptake of isotopes by bottom deposits, by water plants, algae, and fish.

1.3.3 Physiological Types of Pollution

Taste. Industrial effluents contain many chemicals that impart unpleasant tastes to water (e.g., phenols, oil refinery wastes) and may make the water of the receiving streams either unfit for drinking purposes or else difficult and costly to purify.

Odor. Most unpleasant smells associated with polluted rivers are due to the presence of inorganic and organic compounds of nitrogen, sulfur, and phosphorus and arise from the putrefaction of proteins and other organic materials present in sewage and in industrial wastewaters.

The elimination of odors in drinking water can often be accomplished by the use of a few ppm of activated carbon, which has the additional advantage of removing tastes.

1.3.4 Biological Types of Pollution

Under this heading, we must include pathogenic bacteria, certain fungi, algae, viruses, parasitic worms, and any plants or animals which either multiply excessively in a stream or are otherwise undesirable or harmful.

It is generally assumed that if organisms of the coliform group (normal intestinal bacteria which can easily be detected and counted) cannot be found, then the pathogens must also be absent.

Industrial effluents, as a rule, are free from pathogens; although anthrax bacilli may possibly occur in tannery wastes derived from hides.

1.4 Chemical, Physical, and Biological Parameters of Water Quality

Over 400 chemical, physical, and biological parameters are available to characterize water quality. It would be impractical to apply all the known analytical procedures to a water sample. Obviously, to characterize or define certain properties, a choice of parameters must be made. The selection is based on the needs of the program for which the analysis is being performed. In practice, only a few of the more than 400 available measurements are used with any frequency. The most frequently used tests for characterization of water pollution are tabulated in Table 1.4.

Table 1.4 Tests used to measure water pollution

Nutrient demand	Specific nutrients	Nuisances	Toxicity	Physical characteristics	Bacteriological indicators
Dissolved oxygen (DO)	Nitrogen Ammonia	Sulfide	Cyanide	Solids Dissolved	Coliforms Fecal
	Nitrate	Sulfite	Heavy metals	Settleable	Non-Fecal
Biochemical oxygen demand (BOD)	Nitrite			Floating	
	Organic N	Oil	Pesticides	Suspended	Total bacterial count
Chemical oxygen demand (COD)	Phosphorus Orthophosphate	Detergents		Turbidity	
	Polyphosphate Organic P	Phenols		Color	
Total organic carbon (TOC)				Taste Odor Radioactivity	

1.4.1 Analysis of Water and Wastewater

1. Organic nitrogen is determined by the Kjeldahl method. The sample is boiled to remove ammonia and then digested with sulfuric acid: the organic nitrogen is converted to ammonia. The NH_3 content is determined by colorimetric method.
2. Phosphorus: Orthophosphates are determined by addition of ammonium molybdate that forms a colored complex with orthophosphates. The polyphosphates and organic phosphorus need to be converted to orthophosphate before analysis.
3. Turbidity is measured by using a Jackson turbidimeter. The test is based on measuring the length of the light path through which the flame of a standard candle becomes indistinguishable from its background. The units are JTU.
4. Solids content:

The various categories of solids contained in wastewater are listed in Table 1.5.

1.4.2 Dissolved Oxygen and Solids Content in Wastewater

The term "solids" as applied to wastewater is by tradition taken to mean floating solids as well as relatively low boiling point (volatile) soluble compounds. The detailed definitions are in this section.

Table 1.5 gives a summary of the methodology used in the laboratory to obtain experimental values of different types of "solids content" in wastewater [1]:

Table 1.5 Solids content in wastewater

	Sample		
	Total solids (residue 103 °C)	Inorganic (residue 550 °C)	Organic (loss 550 °C)
Unfiltered (suspended + dissolved)	Totals solids (TS)	Total fixed solids	Total volatile solids
Filtered (dissolved)	Total dissolved solids (TDS)	Fixed dissolved solids	Volatile dissolved solids
By difference	Suspended solids (SS)		Volatile suspended solids (VSS)

Reprinted from "Environmental Science and Engineering", by J. Glynn Henry and Gary W. Heinke, Prentice Hall, Upper Saddle River, N.J. 1996, by permission from Pearson Prentice Hall, Upper Saddle River, N.J.

1. The sample of wastewater when unfiltered and heated to 103 °C leaves a residue. This residue from the unfiltered solids is called "*Total Solids*" (TS).
2. When the sample of wastewater is filtered and the filtered water is heated to 103 °C, the residue is called "*Total Dissolved Solids*" (TDS).
3. Suspended Solids (SS) = Total Solids − Total Dissolved Solids = TS − TDS.
4. When the sample of filtered wastewater is heated to 550 °C, the residue on the dish is equal to "*Total Fixed Solids.*"
5. When the sample of wastewater is filtered and then heated to 105 °C, the residue on the dish is equal to "Fixed Dissolved Solids."
6. The "Total Volatile Solids" = "Total Solids" − "Total Fixed Solids."
7. The "Volatile Dissolved Solids" = "Total Dissolved Solids" − "Fixed Dissolved Solids."
8. The "Volatile Suspended Solids" (VSS) = "Total Volatile Solids" − "Volatile Dissolved Solids."

Dissolved Oxygen (DO) is often the most important parameter in studies of pollution. Severe depletion of oxygen concentration may indicate an excessive load of organic wastes. Low dissolved oxygen will have an adverse effect on the fish and other aquatic life of a body of water.

The concentration of DO in water at equilibrium with the atmosphere is a function of temperature. The solubility of oxygen decreases from 14.5 mg/l at 0 °C to 8.4 mg/l at 25 °C in pure water. Uncontaminated surface waters are usually nearly saturated with oxygen. Available nutrients such as carbon, hydrogen, nitrogen, sulfur, and phosphorus compounds encourage microbiological activity with concurrent oxygen depletion. Algal activity may raise the oxygen level during daylight but depress DO during night. DO levels may be used as an indicator of water pollution by oxygen demanding wastes. Low DO concentrations are likely to be associated with low-quality waters.

The difference between the initial and final DO content of a sample of wastewater under specified conditions is called oxygen demand.

Fig. 1.10 a Complete BOD
curve showing the
carbonaceous and
nitrification demands.
b Oxygen utilized versus time

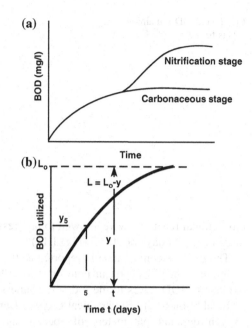

The oxygen demand of polluted waters is exerted by three classes of materials:

1. Carbonated organic material available as a food for micro-organisms.
2. Nitrogenous material susceptible to microbial oxidation, and
3. Chemical reducing agents susceptible to high-rate chemical oxidation, such as ferrous iron, sulfite, sulfide, and certain organic reducing agents.

All three classes are important in the deoxygenation pattern.

Biochemical Oxygen Demand (BOD) is defined as the amount of oxygen (expressed in parts per million, ppm, or in milligrams per liter, mg/l) required for biochemical oxidation of organic and inorganic matter in aerobic bacterial action in an incubated sample of wastewater.

When the oxygen demand of a sewage effluent is measured over a long period of time, for instance, in a series of BOD bottles, the O_2 consumed will vary in accordance with a curve similar to that shown in Fig. 1.10.

The oxidation at first proceeds quite rapidly and then gradually slows down over a period of about 20 days. Subsequently, however, two further stages in the oxidation occur, which often account for a considerable portion of the total O_2 consumption. The demand exerted in the first 20 days is usually mainly due to the oxidation of carbonaceous organic matter and is referred to as the Ultimate First Stage BOD. The later stages are due to the oxidation of nitrogenous material, mainly ammonia to nitrite and ultimately to nitrate. The oxygen demand exerted during the complete oxidation of nitrogenous materials (nitrification) is referred to as Ultimate Second Stage BOD. The Ultimate BOD would be obtained if all the

Fig. 1.11 BOD remaining
versus time

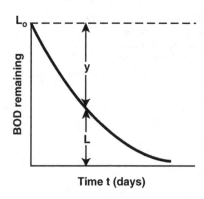

biochemical reactions were allowed to progress to completion. This would require more than 100 days at 20 °C (Fig. 1.11).

The BOD is conventionally reported as the 5 day BOD (BOD_5) and is defined as the amount of oxygen in ppm or mg/l utilized during 5 days of biochemical oxidation. BOD refers to the 5 day test unless specified otherwise.

In addition to the Biochemical Oxygen Demands, there are a number of other oxygen demand parameters of special significance in wastewater treatment (Fig. 1.12).

Immediate Oxygen Demand (IOD) is the depletion of DO in a standard water dilution of the wastewater sample in 15 min. It may include oxygen utilized in reaction with ferrous iron, sulfites, sulfides, or other reducing agents which will react within 15 min of contact.

Chemical Oxygen Demand (COD) measures the total oxidizable material content, both organic and inorganic, as well as biodegradable or non-degradable. It is defined as the O equivalent expressed in ppm or mg/l of the amount of a strong chemical oxidizing agent other than oxygen (usually potassium dichromate in sulfuric acid solution) consumed by oxidizable organic and inorganic material in a sample of wastewater.

For certain wastewaters, a definite COD/BOD relationship may exist. This is true for domestic sewage for which COD may approximate Ultimate First Stage BOD. A high COD/BOD ratio of an organic waste which, from all know characteristics, should be biodegradable, may indicate the presence of toxic materials. A high COD/BOD ratio may also indicate that the waste is resistant to biodegradation.

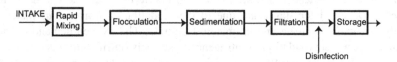

Fig. 1.12 Water treatment plant

Although COD usually exceeds BOD, the reverse may be true with certain wastes, notably cellulose. In such cases, COD is not a reliable measure of deoxygenation potential.

Total Organic Carbon (TOC). Another means for measuring the organic matter present in wastewater is the TOC test, which is especially applicable to small concentrations of organic matter. The test is performed by injecting a known quantity of sample into a high-temperature furnace. The organic carbon is oxidized to carbon dioxide in the presence of a catalyst. The CO_2 that is produced is quantitatively measured by means of an infrared analyzer. Acidification and aeration of the sample prior to analysis eliminates errors due to the presence of inorganic carbon. The test can be performed very rapidly and is becoming more popular.

Mathematical Basis of the BOD Test. Many attempts have been made to express the variation of BOD with time in mathematical terms. Early investigators assumed that the biochemical oxidation is a first-order reaction so that at any instant the rate of oxygen depletion is proportional to the quantity still required for complete oxidation. This may be written as:

$$-\frac{dL}{dt} = K_1 L \tag{1.30}$$

where L is the Ultimate First Stage BOD remaining at any time t, and K is the rate coefficient of oxidation. If we denote the initial value of L by L_O (in other words, L_O represent the total Ultimate First Stage (BOD), then after integration, we obtain:

$$\int_{L_O}^{L} \frac{dL}{L} = -K_1 \int_{o}^{t} dt \tag{1.31}$$

$$\ln \frac{L}{L_O} = -K_1 t \tag{1.32}$$

$$L = L_O \exp(-Kt) \tag{1.33}$$

Sometimes it is more convenient to use the oxygen demand satisfied at time t. Denoting this by y, where

$$y = L_O - L \tag{1.34}$$

Substituting the value of L obtained above:

$$y = L_O\left(1 - e^{-K_1 t}\right) \tag{1.35}$$

Alternately: $y = L_O\left(1 - 10^{-Kt}\right)$, where K is now the rate coefficient to the base 10 and thus numerically $K = \frac{K_1}{2.3}$. Both K and L_O are unknowns in the BOD equations. If K is known, L_O can be calculated from the 5 day BOD results.

1.4.3 Drinking Water and Wastewater Treatment

A block diagram for drinking water treatment is given in Fig. 1.12.

The operations shown in the block diagram are described below. Since the operation of sedimentation/clarification is common to both water treatment and wastewater treatment, the design issues for both of them are given in the section that immediately follows this one.

1.4.3.1 Sedimentation: Theoretical Settling Velocities of Particles

Sedimentation is the process of removal of solids by means of gravity settling. Depending upon the situation, the solids can settle down in the form of either individual particles or agglomerated particles. In this section, the formula for settling velocities of individual particles are presented.

Individual Particles

The force F_D is the drag or frictional force due to the relative motion between the particle and fluid. The drag coefficient C_D is defined as:

$$C_D = \frac{F_D}{A_P \frac{\rho_f V_d^2}{2}} \tag{1.36}$$

A_P Projected area of solid $= \frac{\pi d_p^2}{4}$ for a spherical solid
d_P Diameter of particle
ρ_p Density of particle
ρ_f Density of fluid
V_d Relative velocity between solid and fluid, or terminal settling velocity

Dimensionless Reynolds number Re $= \frac{d_P V_d \rho_f}{\mu_f}$

$\mu_f =$ viscosity of fluid
For water,
$\mu_f = \mu_w$, viscosity of water
The value of C_D is given by

$$C_D = {}^{24}\!/_{Re} \text{ (Laminar; Re} < 1.0)$$
$$= {}^{24}\!/_{Re} + {}^{3}\!/_{(Re)^{1/2}} + 034 \text{ (Transitional)}$$
$$= 0.4 \text{ (Turbulent; Re} \geq 10^4)$$

For a spherical particle, the value of the terminal velocity v_t is given by

Table 1.6 Values of parameters to compute friction factor for solid–fluid flow

Range	Range for Re	b_1	n
Laminar	$0 < Re < 2$	24	1
Transitional	$2 > Re < 500$	18.5	0.6
Turbulent	$Re > 500$	0.44	0
Length: width ratio			
Clarifier	3:1 to 5:1		
Filter Bag	1.2:1 to 1.5:1		
Chlorine contact chamber	20:1 to 50:1		

$$v_t = \sqrt{\frac{4/3\left(\rho_P - \rho_f\right)gd}{C_D\rho_f}} \text{ for all regions - laminar, transitional, and turbulent}$$

$$(1.37)$$

Since the value of C_D in turn depends upon the velocity, one may need to perform a trial and error solution to obtain the value of the velocity. Alternatively, one can use the method advanced by McCabe-Smith [6] by using a different (though similar) formula for C_D:

$$C_D = \frac{b_1}{Re^n} \tag{1.38}$$

where values of b_1 and n are given in the Table 1.6.

The terminal velocity is computed using the formula

$$V_t = \left[\frac{4gd^{(n+1)}\left(\rho_p - \rho_f\right)}{3b_1\mu^n\rho_f^{1-n}}\right]^{\frac{1}{2-n}} \tag{1.39}$$

In order to determine the range of the flow, McCabe-Smith define a K-factor which does not require the velocity for its computation:

$$K = d\frac{\left[g\rho_f\left(\rho_p - \rho_f\right)\right]^{\frac{1}{3}}}{\mu_f^2} \tag{1.40}$$

Then, the problem of determining V_t can be solved by first computing the value of K, then determining the regime to obtain b_1 and n, and finally computing V_t (Table 1.6).

Clarification

There are two definitions of the term clarification.

Clarification is the general term ("make clear") for the combination of the three unit operations: coagulation, flocculation and sedimentation. Coagulation combined with flocculation is the process of neutralization of small charged particles so that they can be brought together into "flocs" through the process of flocculation. Common coagulants are aluminum and ferric sulfates and chlorides, sodium aluminate, etc.

Table 1.7 Typical primary clarifier percent removal

	Overflow rates			
(gpd/ft^2)	1,200	1,000	800	600
(m/d)	48.9	40.7	32.6	24.4
Suspended solids	54 %	58 %	64 %	68 %
BOD$_5$	30 %	32 %	34 %	36 %

Sedimentation is the process of settling the large particles, done using laboratory experimentation, either as they exist or obtained through the process of coagulation and flocculation.

Clarification or sedimentation is an operation that is used for the removal of settleable solids in:

Wastewater treatment: Raw wastewater, wastewater after secondary treatment (activated sludge process or fixed-film reactors), wastewater after chemical precipitation [10].

Water treatment: Raw water to remove settleable solids; to remove flocs after coagulation and flocculation and to remove solids after lime-soda softening for remove hard water.

The velocity of individual particles has been presented in an earlier section. For the design of sedimentation tanks for water treatment or wastewater treatment, the methodology and criteria are given below:

The surface-settling rate (overflow rate) is computed by using incoming flow rate for primary settlers (Table 1.7). For settlers following secondary treatment, strong recirculation patterns within the tank cause influent and effluent flow rates to be different. The effluent flow rate is used to compute the overflow rate.

$$\text{Overflow rate } V_o = \frac{Q}{A_{\text{SURFACE}}} \tag{1.41}$$

Q Average daily flow rate, gallons per day
A_{SURFACE} Total surface area of tank
V_o Overflow rate, gallons per day per ft^2

The horizontal velocity V_n is given by

$$V_n = \frac{Q}{A_{\text{CROSS-SECTION}}} = \frac{Q}{A_x} \tag{1.42}$$

Q average daily flow rate, gallons per day
$A_{\text{cross-section}}$ A_x = cross-section area on the side, ft^2

Table 1.8 Design data for clarifiers for activated sludge systems

Type of treatment	Overflow rate, $m^3/m^2 - d$		Loading $Kg/m^2 - h$		Depth (m)
	Avg	Peak	Avg	Peak	
Settling following air activated sludge (excluding extended aeration)	16–32	40–48	3.0–3.6	9.0	3.5–5
Settling following extended aeration	8–I6	24–32	1.0–5.0	7.0	3.5–5

Reprinted from "Wastewater Engineering: Treatment, Disposal and Reuse", by Tchobanoglous, George and Burton, Franklin, 1991 by permission from McGraw Hill Education, LLC.

Recommended horizontal velocities are given below, as are the recommended dimensions of rectangular and circular tanks (Table 1.8).

Horizontal Velocities

1. Water treatment—horizontal velocities should not exceed 0.5 fpm.
2. Wastewater treatment—no specific requirements (use the same criteria as for water).

Dimensions

1. Rectangular tanks
(a) Length: Width ratio = 3:1 to 5:1
(b) Basic width is determined by the scraper width (or multiples of the scraper width).
(c) Bottom slope is set at 1 %.
(d) Minimum depth is 10 ft.

2. Circular tanks
a. Diameters up to 200 ft.
b. Diameters must match the dimensions of the sludge scraping mechanism.
c. Bottom slope is less than 8 %.
d. Minimum depth is 10 ft.

Effluent weir loading is equal to the effluent flow rate divided by the length of the weir.

Weir overflow rate = WOR = Q/weir length

Recommended weir loading rates are given below, followed by design criteria for primary clarifiers, design data for clarifiers for activated sludge systems, and design criteria for sedimentation basins (Table 1.9).

Weir Loadings

1. Water treatment—weir overflow rates should not exceed 20,000 gpd/ft.
2. Wastewater treatment

(a) Flow < IMGD: weir overflow rates should not exceed 10,000 gpd/ft.
(b) Flow > 1MGD: weir overflow rates should not exceed 15,000 gpd/ft.

Table 1.9 Design criteria for sedimentation basins

Type of basin	Overflow rates (gpd/ft^2)	Detention time (hr)
Water treatment		
Presedimentation	300–500	3–4
Clarification following coagulation and flocculation		
1. Alum coagulation	350–550	4–8
2. Ferric coagulation	550–700	4–8
3. Upflow clarifiers		
A. Ground water	1,500–2,200	1
B. Surface water	1,000–1,500	4
Clarification following lime-soda softening		
1. Conventional	550–1,000	2–4
2. Upflow clarifiers		
a. Ground water	1,000–2,500	1
b. Surface water	1,000–1,800	4
Wastewater treatment		
Primary clarifiers	600–1,200	2
Fixed film reactors		
1. Intermediate and final clarifiers	400–800	2
Activated sludge	800–1,200	2
Chemical precipitation	800–1,200	2

Reprinted from, "FE Fundamentals of Engineering, Supplied- Reference Handbook", NCEES 8th edition, 2nd revision, 2011 by permission from the National Council of Examiners for Engineering and Surveying.

1.4.3.2 Hardness [11]

Hardness is a property of water that prevents lathering of soap. Soap gives off precipitates instead of lather when used with hard water.

Hardness is caused by the *cations* calcium (Ca^{2+}) and magnesium (Mg^{2+}) and to a lesser extent by the uncommon cations iron (Fe^{2+}), manganese (Mn^{2+}), and strontium (Sr^{2+}).

The combination of hardness-causing ions, as well as chemical substances, is given in terms of an equivalent amount of $CaCO_3$ (mol. wt 100, equivalent wt 50).

The equivalent of $CaCO_3$ of a hardness-causing substance or ion is:

= mass of hardness producing substance or ion $\times$ 50/eqvt wt of hardness producing substance or ion.

A table of multiplication factors for several common ions or substances is given in Table 1.10.

The units of hardness are:

1. mg/L of the substance in eqvts of $CaCO_3$.
2. ppm of the substance in terms of equivalent $CaCO_3$.
3. milliequivalents per liter (meq/L) of hardness.

Table 1.10 Multiplication factors for ions

Ion or substance	Molecular weight	Equivalent weight	Multiplication factor[*]
$CaCO_3$	100	50	1
$MgCO_3$	84	42	50/42
$CaSO_4$	136	86	50/68
$CaCl_2$	111	55.5	50/55.5
$Mg(NO_3)_2$	148	74	50/74
		4	
$Ca(HCO_3)_2$	162	81	50/81
$Mg(HCO_3)_2$	146	73	50/73
HCO_3^-	61	61	50/61
CO_3^{2-}	60	30	50/30
OH^-	17	17	50/17
SO_4^{2-}	96	48	50/48

[*] = eqvt wt of $CaCO_3$ (i.e., 50)/eqvt wt of hardness producing substance or ion

$$1 \text{ meq/L} = 50 \text{ ppm } CaCO_3$$

Hardness characterized by water having only $Ca(HCO_3)_2$ and $Mg(HCO_3)_2$ is called temporary hard water. The temporary hard water can be softened by boiling, which leads to the following reactions:

$$Ca(HCO_3)_2 \rightarrow CaCO_3 + H_2O + CO_2 \uparrow$$

$$Mg(HCO_3)_2 \rightarrow Mg(OH)_2 + 2CO_2 \uparrow$$

1.4.3.3 Softening of Hard Water

1. **Lime-soda ash process**. Both lime ($Ca(OH)_2$) and soda ash (Na_2CO_3) are used for the remediation of both carbonate and non-carbonate based hardness from Ca^{2+} and Mg^{2+} ions. It can be carried out at room temperature or at elevated temperatures [9, 10, 15].

   Lime-Soda Softening Equations (Equivalent weights for common compounds and ions are given in Table 1.11.)

(a) Carbon dioxide removal

$$CO_2 + Ca(OH)_2 \rightarrow CaCO_3(s) + H_2O$$

(b) Calcium carbonate hardness removal

$$Ca(HCO_3)_2 + Ca(OH)_2 \rightarrow 2CaCO_3(s) + 2H_2O$$

(c) Calcium non-carbonate hardness removal

$$CaSO_4 + Na_2CO_3 \rightarrow CaCO_3(s) + 2Na^+ + SO_4{}^{2-}$$

(d) Magnesium carbonate hardness removal

$$Mg(HCO_3)_2 + 2Ca(OH)_2 \rightarrow 2CaCO_3(s) + Mg(OH)_2(s) + 2H_2O$$

(e) Magnesium non-carbonate hardness removal

$$MgSO_4 + Ca(OH)_2 + Na_2CO_3(s) \rightarrow CaCO_3(s) + Mg(OH)_2(s) + 2Na^+ \\ + SO4^{2-}$$

(f) Destruction of excess alkalinity

$$2HCO_3^- + Ca(OH)_2 \rightarrow CaCO_3(s) + CO_3^{2-} + 2H_2O$$

(g) Recarbonation

$$Ca^{2+} + 2OH^- + CO_2 \rightarrow CaCO_3(s) + H_2O$$

2. Ion-exchange process. Zeolites, which are compounds of sodium represented by Na_2Ze, are used. Zeolites are either natural (non-porous) or synthetic (porous). Hard water is percolated through a bed of zeolite and the Ca^{2+} and Mg^{2+} ions from the water are exchanged for the Na^+ ions. The isotherm used for describing the ion exchange is a linear isotherm, showing a linear relation between the concentration in the adsorbent and the concentration in the solution [7].

1.4.3.4 Coagulation and Flocculation [2]

Water contains particles of varying sizes. Sedimentation is effective in removing waste and fine particles. However, colloidal solids (sizes 1 to 1000 nm), which cause turbidity in water, are not removed by sedimentation alone, no matter what the detention times and overflow rates. Colloidal particles also generally have an electric charge on their surface and the surface area to volume ratio is high due to their small size.

Coagulation refers to the process of bringing the colloidal particles together. Colloids can be described as either thermodynamically stable (e.g., soaps, detergents, etc.) or unstable (e.g., microorganisms, metal oxides, etc.). Coagulation is useful for unstable colloids. Coagulation can be brought about by the addition of

Table 1.11 Equivalent weights for common compounds

Equivalent weights	Molecular weight	# Equiv mole	Equivalent weight
CO_3^{2-}	60.008	2	30.004
CO_2	44.009	2	22.004
$Ca(OH)_2$	74.092	2	7.046
$CaCO_3$	100.086	2	50.043
$Ca(HCO_3)_2$	162.110	2	81.055
$CaSO_4$	136.104	2	68.070
Ca^{2+}	40.078	2	20.039
H^+	1.008	1	7.008
HCO_3^-	61.016	1	61.016
$Mg(HCO_3)_2$	146.337	2	73.168
$Mg(OH)_2$	58.319	2	29.159
$MgSO_4$	120.367	2	60.184
Mg^{2+}	24.305	2	12.152
Na^+	22.990	1	22.990
Na_2CO_3	105.998	2	52.994
OH^-	17.007	1	17.007
SO_4^{2-}	96.062	2	48.031

coagulant chemicals (e.g., aluminum sulfate) or ion salts. Coagulated particles form "floc" or sticky mixtures which are larger in size and thus can settle down. Synthetic organic chemicals called polyelectrolytes are also added as a coagulant aid along with alum. The combined process of coagulation and flocculation proceeds in two conceptual steps:

1. Charge neutralization when the coagulant chemical neutralizes the charge on the surface of the colloids.
2. Bridging between surfaces to bring together the neutralized particles. Bridging is best brought about when the particles travel at different velocities and thus can collide with each other. This is achieved by a mixing process using flocculator paddles which give a range of velocities to the colloids in the water.

The optimum coagulation chemical dosage can be obtained by means of jar tests. Six beakers are filled with water and differing amounts of the coagulant chemical. The floc, when formed, are allowed to settle. The flocculation is a function of several parameters—pH, alkalinity, temperature and the amount of coagulant chemical. During the jar tests, all of these parameters can be varied if it is desired to perform a complete optimization study. The chemical dosage chosen is the one that uses the least amount of chemical that can perform the task for a given water sample with its given conditions of temperature, pH, and alkalinity. The optimum dosage needs to be found through experimental means.

Flocculation or the process of obtaining "floc" can occur between particles due to two mechanisms:

1. Perikinetic flocculation where the Brownian motion of particles is responsible for flocculation.
2. Orthokinetic flocculation refers to the flocculation due to the artificial motion of the particles, done by mixing. The spatial changes in the velocity, i.e., the velocity gradient G is an important parameter that determines the effectiveness of the flocculation process. The mixing process is described in the next section.

1.4.3.5 Mixing Processes and Filtration

The mixing operation is required for the purpose of dispersing chemicals during the coagulation process [9]. The contact between the chemicals and the suspended particles in the rapid-mix basins leads to the formation of micro-flocs.

Following the formation of microflocs, the water is led to the flocculation basins where large, dense, agglomerated particles are formed.

Mixing is thus an important process in flocculation, mixing of suspensions, and mixing of chemicals. In the continuous rapid mixing process, the power input per unit volume is an approximate measure of the soundness of the process. This is G, the mean velocity gradient of mixing [9, 10]:

$$G = \sqrt{\frac{P}{\mu V}} = \sqrt{\frac{\gamma H_L}{t \mu}} \qquad (1.43)$$

where
$Gt =$ 10^4 to 10^5
G Mixing intensity = root mean square velocity gradient
P Power
V Volume
μ Bulk viscosity
γ Specific weight of water
H_L Head loss of mixing zone, and
t Time in mixing zone

Paddles are gentle mixers and are applied in flocculation processes. The power requirement P_{BOARD} is given by [9, 10]:

Reel and Paddle

$$P_{BOARD} = \frac{C_D A_P \rho_f v_p^3}{2},$$

where
C_D drag coefficient = 1.8 for flat blade with a length:width ratio > 20:1

A_P | Area of blade (m^2) perpendicular to the direction of travel through the water
ρ_f | Density of H_2O (kg/m^3)
v_p | Relative velocity of paddle (m/sec) = V × slip coefficient, and
V | velocity of paddle
Slip coefficient | $0.5 - 0.75$

Propeller and turbine mixers are used in baffled tanks. The baffles reduce the amount of vortexing in the tank; vortex formation results in circulation of the fluid in the tank without any mixing in the perpendicular directions. The power requirement for these types of mixers is given by Reynolds and Richards, and NCEES [9, 10]:

Turbulent Flow Impeller Mixer

$$P = K_T(n)^3 (D_i)^5 \rho_f,$$

where
K_T Impeller constant (see Table 1.12)
n Rotational speed (rev/sec), and
D Impeller diameter (m)

Filtration
Filtration is used in both wastewater and drinking water treatment to remove suspended solids. The pressure drop in a filter bed (packed bed) is due to two reasons:

1. Pressure drop due to the flow of the fluid through channels between particles. This component of the pressure drop is called skin friction.

Table 1.12 Values of the impeller constant K_T (Assume turbulent flow)

Type of impeller	K_T
Propeller, pitch of 1, 3 blades	0.32
Propeller, pitch of 2, 3 blades	1.00
Turbine, 6 flat blades, vaned disk	6.30
Turbine, 6 curved blades	4.80
Fan turbine, 6 blades at 45°	1.65
Shrouded turbine, 6 curved blades	1.08
Shrouded turbine, with stator, no baffles	1.12

Note Constant assumes baffled tanks having four baffles at the tank wall with a width equal to 10 % of the tank diameter

2. Pressure drop due to the abrupt changes in direction of flow of the fluid due to the shape of the particles. Pressure drop is shown in dimensionless terms as the friction factor f'. The Reynolds number Re is the dimensionless velocity. This is called form friction.

Based on the principles stated, there are 2 equations in common use: the Carmen–Kozeny equation and the Rose equation.

For the Carmen–Kozeny equation [6], the relationship between f' and Re is:

$$f' = \frac{150(1 - \eta)}{\text{Re}} + 1.75 \tag{1.45}$$

The first term on the right hand side represents *skin friction*, and the second term on the right hand side represents *form friction*.

The head loss h_f is related to the friction factor f' by means of two equations; the first one for a bed in which all solids in the bed are of equal size; the second one for a bed in which solids are multisized.

Monosized media

$$h_f = \frac{f'L(1 - \eta)V_s^2}{\eta^3 g d_p} \tag{1.46}$$

Multisized media

$$h_f = \frac{L(1 - \eta)V_s^2}{\eta^3 g} \sum \frac{f'_{ij} x_{ij}}{d_{ij}} \tag{1.47}$$

h_f Head loss through the clean bed (m of H_2O)
L Depth of filter media (m)
η Porosity of bed = void volume/total volume,
V_s Filtration rate = empty bed approach velocity

    = Q/A_{plan} (m/s), and
g Gravitational acceleration (m/s^2)

$$\text{Re} = \text{Reynolds number} = \frac{V_s \rho d}{\mu} \tag{1.48}$$

d_{ij} Diameter of filter media particles; arithmetic average of adjacent screen openings (m)
i Filter media (sand, anthracite, garnet)
j Filter media particle size
x_{ij} Mass fraction of media retrained between adjacent sieves
f'_{ij} Friction factors for each media fraction

The alternative equation, the Rose equation [16, 17, 18] is used mainly for computations involving sand filters in which there is a range of particles from top to bottom in the filter. The Rose equation relates the head loss for flow of fluid through the packed bed to the velocity, bed dimensions, solid dimensions, and the drag coefficients for the hypothetical settling of the solids through a fluid medium.

Rose Equation:

Monosized (equal sized) media, typically sand

$$h_f = \frac{1.067(V_s)^2 L C_D}{g \eta^4 d} \tag{1.49}$$

Multisized media—typically sand

$$h_f = \frac{1.067(V_s)^2 L}{g \eta^4} \sum \frac{C_{Dij} x_{ij}}{d_{ij}} \tag{1.50}$$

C_D Drag coefficient as defined in settling velocity equations

After filtration, it is customary to clean the bed with the flow of fresh water.

Cleaning of the bed is done by passing water upwards through the bed in order to flush out the filtered solids. The filter bed expands and behaves as a fluidized bed during the cleaning process. The equations for the expanded bed are:

Bed Expansion

Monosized

$$L_{fb} = \frac{L_0(1 - \eta_0)}{1 - \left(\frac{V_B}{V_t}\right)^{0.22}} \tag{1.51}$$

Multisized

$$L_{fb} = L_0(1 - \eta_0) \sum \frac{x_{ij}}{1 - \left(\frac{V_B}{V_t}\right)^{0.22}} \tag{1.52}$$

$$\eta_{fb} = \left(\frac{V_B}{V_t}\right)^{0.22} \tag{1.53}$$

where

L_{fb} Depth of fluidized filter media (m)
V_B Backwash velocity (m/s), Q/A_{PLAN}
V_t Terminal settling velocity; and
η_{fb} Porosity of fluidized bed
L_0 Initial bed depth
η_0 Initial bed porosity

1.4.3.6 Adsorption, Air Stripping, Flotation, and Wastewater Treatment Plants

The operation of adsorption for removing organic matter from water is used for the removal of residual organic substances. The organic matter in the wastewater is collected at the solid–liquid interface. There are two major formulas in use for computing the adsorptive capacity of activated carbon [7, 9, 19]. The Freundlich isotherm is an empirical formula in which the mass of the organic matter adsorbed per unit mass of adsorbent carbon (i.e., x/m) is given as a function of the equilibrium concentration of the organic matter in the wastewater (i.e., C_e), raised to the power $(1/n)$, where n is an experimental constant.

Freundlich Isotherm

$$\frac{x}{m} = X = KC_e^{1/n} \tag{1.54}$$

Here, K is an experimental constant. The linearized form of the Freundlich isotherm is

$$\ln\frac{x}{m} = \frac{1}{n}\ln C_e + \ln K \tag{1.55}$$

For a linear isotherm, $n = 1$.

The Langmuir adsorption isotherm was derived on the basis of two assumptions: There are a fixed number of sites on the carbon for adsorption, and, at equilibrium, the rate of desorption equals the rate of adsorption.

$$\frac{x}{m} = X = \frac{aKC_e}{1 + KC_e}, \tag{1.56}$$

where
a Mass of adsorbed solute required to completely saturate a unit mass of adsorbent, and
K Experimental constant

Linearized Form

$$\frac{m}{x} = \frac{1}{a} + \frac{1}{aK}\frac{1}{C_e} \tag{1.57}$$

Unsteady-state adsorption in a fixed bed

In the process of fixed bed adsorption, which is an unsteady-state process, water having an organic concentration of C_o is passed through an adsorption column of height Z. In the initial period, the carbon in the bed that is located close to the liquid inlet adsorbs the organic compounds. Generally, the carbon gets saturated and the work of adsorption is done by carbon away from the inlet. At the initial stage, the exit concentration of organics in the water is close to zero which is shown in the figure. When the cumulative volume V_B is reached, the exit concentration is $0.05\,C_o$.

Fig. 1.13 Breakthrough curve for a fixed bed adsorption column

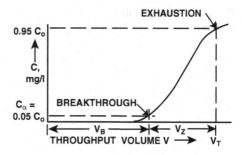

Fig. 1.14 Fixed bed adsorption column

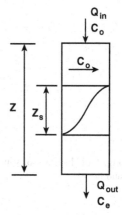

After this point, the S-shaped curve moves up rapidly to 0.95 C_o. The depth of the sorption zone and the breakthrough are shown in Figs. 1.13 and 1.14. It is possible to design a column with laboratory data that gives us S shaped curves that are then arranged to obtain values of V_Z, V_T, V_B, Z_S, and Z.

Depth of Sorption Zone

$$Z_s = Z\left[\frac{V_Z}{V_T - 0.5V_Z}\right] \tag{1.58}$$

V_Z $V_T - V_B$
Z_S Depth of sorption zone
Z Total carbon depth
V_T Total volume treated at exhaustion ($C = 0.95\ C_O$)
V_B Total volume at breakthrough ($C = C_\alpha = 0.05\ C_O$), and
C_O Concentration of contaminant in influent

Air Stripping of Volatiles

Air stripping of wastewater is used for the removal of volatile compounds. Air stripping can be performed either in a packed bed or by means of blowing air past nozzles through which the water (containing the dissolved gases) is made to flow

Fig. 1.15 Packed bed
stripper

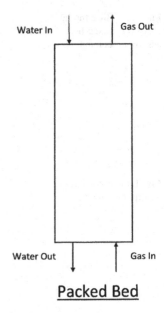

Fig. 1.16 Interface of vapor
and liquid

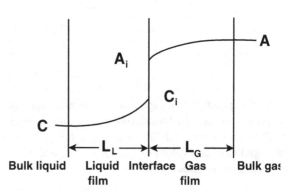

at high velocity. Water is introduced at the top of the packed tower and fresh air is
introduced at the bottom. Compounds removed by air stripping are generally
sparingly soluble in water and thus Henry's Law is used to describe the equilibrium relationship between the concentration in air and in wastewater.

$$A = HC \qquad (1.59)$$

In this equation, P is the partial pressure of the contaminant in the air stream,
H is the Henry's Law constant, and C is the concentration of the contaminant in the
water stream. A diagram of the packed bed is shown in Fig. 1.15.

The mass transfer from the liquid phase to the vapor phase is shown in Fig. 1.16
at any random cross-section of bed. C_i and A_i represent the concentration of the

pollutant in the liquid and gas phases, respectively, at the interface at the given cross-section of the bed. It is assumed that the C_i and A_i are at equilibrium, so that

$$A_i = H'C_i \tag{1.60}$$

In this equation, H' is the dimensionless Henry's Law constant. Since a concentration gradient between C and C_i and also between A and A_i is needed for the mass transfer mechanism to occur and remove the pollutant, an infinitely long packed bed would be needed for completely cleaning up the wastewater, i.e., to obtain $C_{out} = 0$. For the special case of the (hypothetical) infinite packed bed, the outlet concentration in air would be in equilibrium with the incoming concentration in the wastewater, i.e., $A_{out} = H'C_{in}$. The methodology to compute the height of a packed bed needed to accomplish a certain purpose is:

(a) Conduct a material mass balance on the packed bed. In the mass balance the water flow rate Q_w and the incoming concentration is set by the design condition while the outgoing concentration in water, C_{out} is set by the legal requirements.

In general, the mass balance equation is
Removal rate of water = Addition rate to air; assuming that C_{out} and A_{in} are negligible gives

$$Q_W(C_{IN} - C_{OUT}) = Q_W(C_{IN} - 0) = Q_W C_{IN} = Q_A(A_{OUT} - 0) = Q_A A_{OUT} \tag{1.61}$$

Mass transfer through droplets coming out of nozzles for the case of butane-water solutions was studied by Lin [20] and Naimpally [21]. Most of the mass transfer occurs during the breakup of the jet into droplets. Operational simplicity has made packed beds the equipment of choice for removal of sparingly soluble gases from water.

(b) The second step is to compute the minimum air flow needed in order to accomplish the design purpose for the packed bed.

For the special case of an infinitely long bed in which $A_{out} = H'C_{in}$,

$$Q_W C_{IN} = Q_A A_{OUT} = Q_A(H'C_{IN}) \tag{1.62}$$

The stripping factor R is defined as $Q_A H'/Q_W = R$. For an infinitely long packed bed, $R = 1$ and the value of $Q_A = Q_{AMIN}$, which is the minimum air flow rate.

(c) The actual height of the packed bed is based upon the height needed to accomplish the purpose for the case when the actual air flow rate is greater than the hypothetical minimum air flow rate. The design air flow is generally taken to be between $1.5\ Q_{AMIN}$ and $1.75\ Q_{AMIN}$.
  The height Z of the packed bed can be computed, in the general case, by means of an elaborate graphical technique described by Treybal [19] and Seader and Henley [22]. The design is based upon graphical manipulations performed

using two lines representing the equilibrium curve for the specific volatile component and the material balance equation described earlier, both of which are plotted on a graph of mole fraction of the volatile gas in air versus mole fraction of the volatile component in water. For the case of sparingly soluble gases, the simplified algebraic method of NTU-HTU can be used:

For the case of a packed bed of finite height Z, the mass flow rate of air A is greater than the minimum calculated above. Then, the value of the height Z can be calculated using the formula given below for the case of a dilute solution.

The number of transfer units (NTU) multiplied by the height of a transfer unit (HTU) gives the height of the column Z. The computation of NTU can be done by the Colburn [7, 19, 22] equation where R is the stripping factor.

$$\text{NTU} = \text{number of transfer units} = \left(\frac{R}{R-1}\right) \ln\left(\frac{(C_{IN}/C_{OUT})(R-1)+1}{R}\right) \quad (1.63)$$

$$\text{HTU} = \frac{L}{M_W K_L a} \quad (1.64)$$

In this equation, L is the liquid molar loading rate, M_W is the molar density of water, and $K_L a$ (the product of the overall liquid phase mass transfer coefficient and the specific interfacial area) is the transfer rate constant in units of inverse time.

$$Z = (\text{HTU})(\text{NTU})$$

Flotation

Flotation is an operation used in water treatment and wastewater treatment to removed suspended solids. The methodology is to introduce gas bubbles in the liquid phase. These bubbles rise to the top and carry particles with them to the top of the liquid. There are two techniques for carrying it out.

1. Dispersed air technique where the air is introduced through a sparging or bubbling process.
2. Dissolved air process where the precipitated foam is a supersaturated solution.

Flotation is used in conjunction with:

1. Sedimentation.
2. Concentration of biological sludges.

Wastewater Treatment Plants (Fig. 1.17)

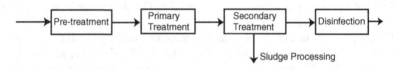

Fig. 1.17 Wastewater treatment plant

Pretreatment consists of:

1. Screening to remove solids.
2. Creation of a constant flow input into the rest of the process by equalization of flow through intermediate storage.
3. Neutralization, where acidic and alkaline streams are mixed or, if these are not available, using acids to neutralize alkaline streams and vice versa.

Primary Treatment consists of the process of sedimentation to remove solids by settling. In the earlier section on "Water Treatment," the process of settling for individual particles was described; also covered was the section on design criteria for settling tanks as applied to drinking water treatment and wastewater treatment. The other categories of settling are flocculant settling and zone settling, which are described below.

Flocculant settling occurs when the settling velocity of the particle increases as it settles through the tank depth because of coalescence with other particles. This increases the settling rate, yielding a curvilinear settling path. For discrete settling, the efficiency of removal is related only to the overflow rate, but when flocculation occurs, both overflow rate and detention time become significant.

Since a mathematical analysis is not easy in the case of flocculant suspensions, a laboratory settling analysis is required to establish the necessary parameters. The laboratory study can be conducted in a column of the type shown in the Fig. 1.18. Suspended solids are determined on the samples drawn off from taps located at 2 ft depth increments at selected time intervals. The data collected from the 2-, 4-, and 6-ft depth taps are used to develop the settling rate versus time relationships (Fig. 1.19).

The results obtained are expressed in terms of percent removal of suspended solids at each tap and time interval. These percentage removals are then plotted against their respective times and depths, as shown in the example that follows. Then, smooth curves are drawn connecting points of equal removal.

The overflow rate is $\frac{Q}{A} = v_o$

In this equation, v_0 is the terminal settling velocity of smallest particle that is 100% removed, Q is the flow rate, and A is the surface area. All particles having a

Fig. 1.18 Taps in a jar

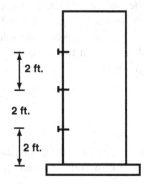

2 ft.

2 ft.

2 ft.

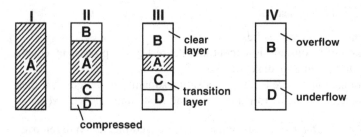

Fig. 1.19 Settling process

settling velocity equal to or greater than v_o will be removed in an ideal settling basin having an overflow rate of $\frac{Q}{A} = v_o$. Particles with a lesser settling velocity v will be removed in the proportion $\frac{v}{v_o}$.

Zone settling involves a flocculated suspension in which the floc particles adhere together and settle as a mass, forming a distinct interface between the floc and the supernatant liquid.

The settling process takes place in three zones, as shown in Fig. 1.19. Initially, all the sludge is at a uniform concentration (A in the Fig. 1.19).

During the initial settling period, the suspended solids settle at a uniform velocity. The settling rate is a function of the initial solids concentration A. As settling proceeds, a volume of relatively clear water B is produced above the zone settling region. Particles remaining in this region settle as discrete particles as discussed previously. A distinct interface exists between the discrete settling region B and the hindered or zone settling region A.

As settling continues, a compressed layer of particles D begins to form at the bottom of the settling unit in the compression zone.

Settling tests are usually required to determine the settling characteristics of the suspension when zone settling and compression are important considerations.

Zone settling rate is lower than the rate for individual particles and thus is used for design.

The area requirement for zone settling is determined as follows: A column of height H_o is filled with a zone settling suspension of uniform concentration C_o. As the suspension settles, the position of the interface as time elapses will be plotted as shown in Fig. 1.20a and b.

The rate at which the interface subsides is then equal to the slope of the curve at that point in time. According to this procedure, the critical area for thickening is given by:

$$A = \frac{Q t_u}{H_o} \tag{1.66}$$

where
A Area required for sludge thickening, ft^2
Q Flow rate into tank, cfs

Fig. 1.20 a Settling curve.
b Computations using settling
curve

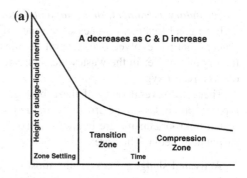

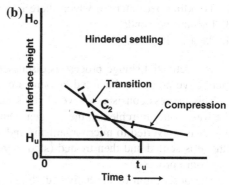

H_o Initial height of interface in column, ft
t_u Time to reach desired underflow concentration, sec

The critical concentration controlling the sludge-handling capability of the tank occurs at a height H_2 where concentration is C_2. This point is determined by extending the tangents to the discrete settling and compression regions of the subsidence curve to the point of intersection and bisecting the angle so formed, as shown in the Fig. 1.20. The time t_u can be determined as follows:

1. Construct a horizontal line at the depth H_u that corresponds to the depth at which all the solids are at the desired underflow concentration C_u. Thus

$$H_u = \frac{C_o H_o}{C_u} \tag{1.67}$$

2. Construct a tangent to the settling curve at the point indicated by C_2.
3. Construct a vertical line from the point of intersection of these two lines to the time axis to determine t_u.

Secondary Treatment, or Biological Treatment Process

Biological treatment processes are part of the traditional wastewater treatment process that has evolved over several decades. The typical process is one in which the organic matter in the wastewater is treated by microorganisms in the presence of dissolved oxygen.

There are several types of these biological treatment processes. The different types of processes can be broadly classified as being either based on fixed media in which microorganisms are bound to a fixed solid base, the trickling bed filter, or suspended media. They are:

1. Activated sludge processes, the most common ones.
2. Trickling bed filters in which there is a fixed media.
3. Facultative ponds.
4. Aerated lagoons.

The activated sludge process is discussed in this section. Trickling bed filters, facultative ponds, and aerated lagoons are discussed in the next section.

This process consists of several steps beginning with the artificial aeration of the wastewater to replenish the oxygen. During the treatment process, the organic matter is converted to microorganisms and the sludge resulting from the resulting sludge is settled and then treated (sludge treatment) through either the aerobic or anaerobic process.

The description of the activated sludge process in the pages to follow begins with an introduction to the kinetics of bacterial growth. This is followed by the principles of aeration of water and wastewater; aeration is needed to replenish the oxygen consumed by the microorganisms. Finally, the process itself is described.

Bacteria reproduce by dividing: the original cell becomes two new organisms. The time required for each division, the generation time, can vary from days to less than 20 min.

If a small number of organisms are inoculated into a culture medium, and the number of viable organisms is recorded as a function of time, the growth pattern based on the number of cells has four distinct phases [23]:

1. The lag phase represents the time required for the organisms to acclimate to their new environment (Fig. 1.21).
2. The log growth phase. During this period, the cells divide at a rate determined by their generation time and their ability to process food.
3. The stationary phase. The population remains stationary because (a) the cells have depleted the substrate or nutrient necessary for growth, toxic end products of metabolism have built up, or there is insufficient space, and (b) the growth of new cells is offset by the death of old cells.
4. The log death phase. The bacterial death rate exceeds the production of new cells.

Fig. 1.21 Cell growth curve

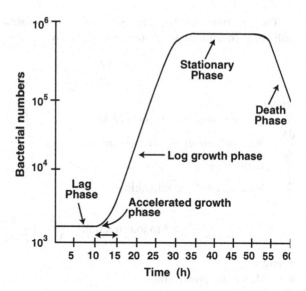

The growth rates are described by two formulas:
Exponential (log) growth with constant specific growth rate, μ

$$\mu = \left(\frac{1}{x}\right)\left(\frac{dx}{dt}\right) \tag{1.68}$$

where

x The cell/organism number or cell/organism concentration

t Time (hr)

μ The specific growth rate (time^{-1}) while in the exponential growth phase

The logistical growth function for batch growth including initial stationary phase is

$$\frac{dx}{dt} = kx\left(1 - \frac{x}{x_\infty}\right) \tag{1.69}$$

$$x = \frac{x_o e^{kt}}{1 - \frac{x_o}{x_\infty}\left(1 - e^{kt}\right)} \tag{1.70}$$

where,

k Logistic growth constant (h^{-1}),

x_o Initial concentration (g/l)

x_∞ Carrying capacity (g/l)

The equations for the rates of growth of microorganisms as a function of the utilization of a limiting substrate are given by:

$$\frac{dX}{dt} = -Y\frac{ds}{dt} \qquad (1.71)$$

where

X   Concentration of microorganisms

$\dfrac{dX}{dt}$ Rate of growth of microorganisms

Y   Growth yield

$\dfrac{ds}{dt}$ Rate of substrate utilization

$$\text{Monod function} = \frac{dX}{Xdt} = \frac{ks}{K_s + s} \qquad (1.72)$$

where

k   Maximum growth rate

K_s the half-saturation constant

s   Concentration of limiting substrate

Aeration of water and wastewater

Gas molecules such as oxygen molecules are transferred from the gas phase to the liquid phase by a three step process [19]:

1. Diffusion through the gas phase to arrive at the interface.
2. Transport across the interface.
3. Transport into the bulk liquid phase.

For oxygen, the liquid film at the interface is the largest resistance. The liquid-phase mass transfer coefficient is defined by:

$$\frac{dC_L}{dt} = K_L a(C_S - C_L) \qquad (1.73)$$

where,

C_L Concentration of oxygen in the liquid phase

C_S Concentration of oxygen in liquid phase at equilibrium with concentration of oxygen in gas phase (saturation concentration)

K_L Volumetric overall mass transfer coefficient

Here, a = area of interface between the gas and liquid, per unit volume of liquid.

The value of $K_L a$ is generally given as a composite quantity.

There are four factors that affect the oxygen transfer rate: the saturated level of oxygen in the liquid, the temperature of the system, the type of waste, and the type of aeration device [23].

Fig. 1.22 Simple activated sludge process

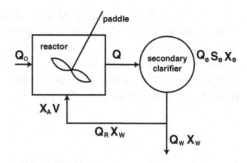

The saturated level of oxygen concentration in the liquid, C_S, is a function of:

(a) Temperature: Tables of C_S versus T are available and C_S decreases with T.
(b) Pressure: $C_S = hp$ where h is a constant and p is the pressure.
(c) Depth of liquid: the saturation level increases with the depth of the tank.

The value of $K_L a$ is a function of temperature and is given as:

$$K_L a|_{T °C} = (K_L a|_{20 °C})(1.20)^{T-20} \qquad (1.74)$$

where T: temperature, °C

The effect of the type of aeration device and type of waste on $K_L a$ needs to be determined by experimentation. The values of α need to be obtained and tabulated, where

$$\alpha = \frac{K_L a \text{ of wastewater}}{K_L a \text{ of plain water}} \qquad (1.75)$$

The activated sludge process is a biological treatment process in which a biologically active microorganism is continuously mixed with sewage flow in an aeration region in the presence of oxygen. Activated sludge used in the process is sludge obtained from the treatment of sewage which had previously been treated while being agitated in the presence of oxygen. Some disadvantages of the process are: Clear, efficient, freedom from odors; limited land area. Disadvantages are: skilled personnel needed; no shocks or sudden changes of flow can be applied to the system; large capital investment. The diagram is shown in Fig. 1.22. The parameter used for determining the return sludge is the SVI, the sludge volume index.

$$\text{SVI} = \frac{\text{sludge volume after settling}}{X_{SS}} \times 1,000 \qquad (1.76)$$

$$\text{SVI} = \frac{V}{X_{SS}} \times 1,000 \qquad (1.77)$$

V Volume of settled solids in 1 L graduated cylinder after 30 min, ml/l
X_{SS} Mixed liquor suspended solids, mg/l (also called MLSS)

It is important to maintain a good level of mixed liquor suspended solids.

The activated sludge process in its simplest form, as shown in Fig. 1.22, consists of a mixing/aeration tank following by a settler/clarifier [10, 11, 23]:

Material balance computations for each unit for each of the components, along with the use of the equations for the growth of microorganisms and the substrate yield the following equations. The biomass concentration in the aeration tank is

$$X_A = \frac{\theta_C Y (S_0 - S_e)}{\theta (1 + k_d \theta_C)} \tag{1.78}$$

S_0 and S_e are the influent and effluent BOD or COD concentration, respectively, and k_d is the microbial death ratio. The yield coefficient is

$$Y = \frac{\text{mass of biomass}}{\text{mass of BOD consumed}} \tag{1.79}$$

The hydraulic residence time is

$$\theta = \frac{V}{Q} \tag{1.80}$$

V is the volume of the aeration basin. The solids residence time is

$$\theta_C = \frac{V X_A}{Q_W X_W + Q_e X_e} \tag{1.81}$$

Q_W and Q_e are the waste sludge and effluent flow rates, respectively, and X_W and X_e are the waste sludge and effluent suspended solids concentrations, respectively. The sludge volume per day, where M is the sludge production rate on a dry weight basis and ρ_S is the wet sludge density, is

$$Q_S = \frac{100\,M}{\rho_S(\text{percentage of solids})} \tag{1.82}$$

$$\text{solids loading rate } = Q X_A \tag{1.83}$$

For an activated sludge secondary clarifier, Q_0 and Q_R are the influent and recycle flow rates, respectively, and

$$Q = Q_0 + Q_R \tag{1.84}$$

$$\text{Organic loading rate (volumetric)} = \frac{Q_0 S_0}{V} \tag{1.85}$$

$$\text{Organic loading rate } (F : M) = \frac{Q_0 S_0}{V X_A} \tag{1.86}$$

Organic loading rate,

where A_M is the surface area of media in a fixed-film reactor $= \dfrac{Q_0 S_0}{A_M}$ $\qquad (1.87)$

$$SVI = \frac{\left(\text{sludge volume after settling }_{mL/L}\right) 1,000}{MLSS_{mg/L}} \qquad (1.88)$$

The steady-state mass balance for a secondary clarifier is

$$(Q_0 + Q_R)X_A = Q_e X_e + Q_R X + Q_w X_w \qquad (1.89)$$

The recycle ratio is

$$R = \frac{Q_R}{Q_0} \qquad (1.90)$$

The recycle flow rate is

$$Q_R = Q_0 R \qquad (1.91)$$

Design and operational parameters for activated sludge treatment of municipal wastewater are given in Table 1.13 for different types of activated sludge processes.

Some of the above types of activated sludge processes are described below (Fig. 1.23); the high purity oxygen and high-rate processes require more operational costs [11, 12, 23]:

Conventional flow shown in Fig. 1.23a where the flow goes through a long aerated tank.

Complete Mix. The complete mix process (Fig. 1.23b) represents an attempt to duplicate the hydraulic conditions of a mechanically stirred reactor. The influent sewage and return sludge flow are introduced at several points in the aeration tank from a central channel. The mixed liquor is aerated as it passes from the central channel to the effluent channels. The effluent is then settled in the settling tank.

The organic load on the aeration tank and the oxygen demand are uniform from one end to the other. As the mixed liquor passes across the aeration tank, it is completely mixed by diffused or mechanical aeration.

Step Aeration. The step aeration process is a modification of the activated sludge process in which the influent sewage is introduced at several points, thus lowering the peak oxygen demand (Fig. 1.23c).

The aeration tank is subdivided into four or more parallel channels through the use of baffles. Each channel comprises a separate step, and the several steps linked together in series. Return activated sludge enters the first step along with a portion of influent sewage. The piping is so arranged that an increment of sewage is introduced into the aeration tank at each step.

The multiple point introduction of sewage maintains an activated sludge with high absorptive properties, so that the soluble organics are removed within a relatively short contact period. Higher BOD loading is therefore possible per 1,000 ft^3 of aeration tank volume. Also in step aeration, the oxygen demand is more uniformly spread over the length of the aeration tank, resulting in better utilization of the oxygen supplied.

Table 1.13 Design and operational parameters for activated sludge treatment of municipal wastewater (PF = plug flow; CM = completely mixed)

Type of Process	Mean cell residence time (θ_c, d)	Food-to-mass ratio (kg BOD$_5$/kg MLSS)	Volumetric loading (V_L, kg BOD$_5$/m^3)	Hydraulic retention time in aeration basin (θ, h)	Mixed liquor suspended solids (MLSS, mg/L)	Recycle ratio (Q_r/Q)	Flow regime	BOD5 removal efficiency (%)	Air supplied (m^3/kg BOD$_5$)
Tapered aeration	5–15	0.2–0.4	0.3–0.6	4–8	1,500–3,000	0.25–0.5	PF	85–95	45–90
Conventional	4–15	0.2–0.4	0.3–0.6	4–8	1,500–3,000	0.25–0.5	PF	85–95	45–90
Step aeration	4–15	0.2–0.4	0.6–1.0	3–5	2,000–3,500	0.25–0.75	PF	85–95	45–90
Completely mixed	4–15	0.2–0.4	0.8–2.0	3–5	3,000–6,000	0.25–1.0	CM	85–95	45–90
Contact stabilization	4–15	0.2–0.6	1.0–1.2			0.25–1.0			45–90
Contact basin				0.5–1.0	1,000–3,000		PF	80–90	
Stabilization basin				4–6	4,000–10,000		PF		
High-rate aeration	4–15	0.4–1.5	1.6–16	0.5–2.0	4,000–10,000	1.0–5.0	CM	75–90	25–45
Pure oxygen	8–20	0.2–1.0	1.6–4	1–3	6,000–8,000	0.25–0.5	CM	85–95	
Extended aeration	20–30	0.05–0.15	0.16–0.40	18–24	3,000–6,000	0.75–1.50	CM	75–90	90–125

Reprinted from, "Wastewater Engineering: Treatment, Disposal and Reuse", by Tchobanoglous, George and Burton, Franklin, 1991 by permission from McGraw Hill Education, LLC.

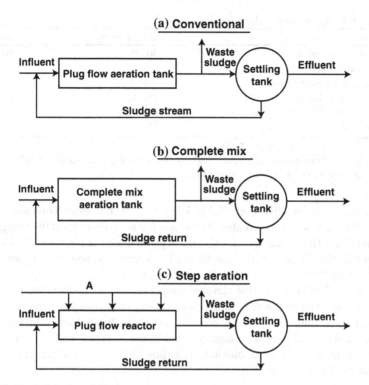

Fig. 1.23 Typical Activated Sludge processes

Contact Stabilization. The contact stabilization process was developed to take advantage of the absorptive properties of activated sludge in the treatment of wastes containing a high proportion of the BOD in suspended or colloidal form.

BOD removal occurs in two stages in the activated sludge process. The first is the absorptive phase which requires 20–30 min. During this phase, most of the colloidal, finely suspended, and dissolved organics are absorbed in the activated sludge. The second phase, oxidation, then occurs and the absorbed organics are metabolically assimilated. In the processes mentioned so far, these two phases occur in a single tank. In the contact stabilization process, the two phases are separated in different tanks.

Sludge digestion

The sludge produced in the activated sludge process is removed, except for the part that is recycled as a part of the process, and is treated further by the process of sludge digestion. The process can be either aerobic or anaerobic [9].

Aerobic digestion is similar to the activated sludge process with similar equipment and tanks. There are fewer operational problems compared to anaerobic digestion although the energy requirements are higher due to the mixing that is required.

Table 1.14 Design parameters for anaerobic digesters

Parameter	Standard-rate	High-rate
Solids retention time, d	30–90	10–20
Volatile solids loading, kg/m^3/d	0.5–1.6	1.6–6.4
Digested solids concentration, %	4–6	4–6
Volatile solids reduction, %	35–50	45–55
Gas production (m^3/kg VSS added)	0.5–0.55	0.6–0.65
Methane content, %	65	65

Source Adapted from Metcalf and Eddy, Inc
Reprinted from, "Environmental Engineering" by Peavy, Howard and Rowe, Donald, 1985 by permission from McGraw Hill Education LLC.

Anaerobic sludge digestion (Table 1.14) [11] is the most commonly used method for the treatment of sludge. The anaerobic bacteria convert the sludge into CO_2 and CH_4. The advantages are that the organic matter is converted into simple compounds; the digested sludge can be used as a fertilizer, and the coliform count is reduced by 99.9 %.

Some disadvantages are that elevated temperature is needed and the process is very sensitive. The mechanism of digestion occurs in two steps: in the first stage, the complex organic compounds, proteins, etc., are attached and converted into organic acids; in the second stage, the acids are attached and converted into gaseous methane and carbon dioxide. Anaerobic digesters are discussed in more detail in the following section.

1.4.3.7 Anaerobic Digesters, Trickling Bed Filters, Facultative Ponds, and Aerated Lagoons [9, 10, 11]

1.   Anaerobic Digesters

Standard Rate or Conventional Digester
In a conventional digester, the digester times are 30 to 60 days, with intermittent feed, mixing and sludge withdrawal. They have either fixed or floating covers. The reactor volume can be computed from the formula:

$$\text{Reactor volume} = \frac{V_1 + V_2}{2} t_r + V_2 t_s \qquad (1.92)$$

where
V_1 Rate of raw sludge input
V_2 Rate of accumulation of digested sludge
t_r Time to react and thicken
t_s Storage time

High-Rate Digesters
High-rate digesters complete their work in 10–20 days of residence time and have continuous mixing, withdrawal and addition of sludge. Two-stage digestion

is the preferred mode: in the first stage, the main action is liquefaction, digestion of soluble organics and gasification. In the second stage, the actions are the separation of liquid, removal of gas and storage of the digested sludge. The formulas for the volume of vessel for each stage are:

First stage

$$\text{Reactor volume} = V_1 t_r \qquad (1.93)$$

Second stage

$$\text{Reactor volume} = \frac{V_1 + V_2}{2} t_r + V_2 t_s, \qquad (1.94)$$

where

V_1 Raw sludge input (volume/day)
V_2 Digested sludge accumulation (volume/day)
t_r Time to react in a high-rate digester = time to react and thicken in a standard-rate digester
t_t Time to thicken in a high-rate digester, and
t_s Storage time

2. Trickling Bed Filter [10, 23]:

A trickling bed filter consists of impervious materials like stone or plastics, on which the sewage is sprayed. The bed like structure needs to have a sufficient amount of ventilation to ensure a fresh supply of oxygen. The sewage flows over a film that is made up of two parts: an aerobic film which is up to a depth of 0.15 mm and an anaerobic film below it. During the flow over the aerobic film, the soluble organic matter is metabolized and the colloidal matter is absorbed on to the surface. The organisms attached to the upper part of the bed grow more, while in the lower part of the bed, the organic content decreases and the microorganisms are deprived of food. In the lower region, the microorganisms utilize their own cytoplasm and thus are in an endogenous growth phase. They then lose their ability to hold on to the solid medium and are washed off. Some of the advantages of trickling bed filters are simplicity of design and operation, low operating costs, and good effluent under widely varying conditions. Some disadvantages are odor and unwelcome smell, head loss of water, large area required, and finally, due to the unsightliness and smell, are not suitable to be located in populated areas.

Recirculation is needed for a high rate of flow. A part of the effluent is pumped back to the inflow, which provides long contact times for organic matter and also high loading without clogging.

The recirculation ratio R is equal to the ratio of the recirculated flow Q_R to the inflow of raw sewage, Q_0.

BioTower

Trickling beds made out of plastic have a relatively more predictable behavior. Several equations have been developed by the National Research Council and Professor W.W. Eckenfelder.

Fixed-Film Equation without Recycle

$$\frac{S_e}{S_o} = e^{-kD/q^n} \tag{1.95}$$

Fixed-Film Equation with Recycle

$$\frac{S_e}{S_a} = \frac{e^{-kD/q^n}}{(1+R) - R(e^{-kD/q^n})} \tag{1.96}$$

$$S_a = \frac{S_o + RS_e}{1 + R} \tag{1.97}$$

S_e Effluent BOD$_5$ (mg/L)
S_o Influent BOD$_5$ (mg/L)
D Depth of biotower media (m)

$$q = \text{Hydraulic loading } (m^3/m^2/min)$$
$$= (Q_o + RQ_o)/A_{plan}(\text{with recycle})$$

k Treatablility constant; functions of wastewater and medium (min^{-1}); range
      $0.01 - 0.1$; for municipal wastewater and modular plastic media 0.06 min^{-1}
      @ 20 °C,
k_T $k_{20}(1.035)^{T-20}$
n Coefficient related to media characteristics; modular plastic $n = 0.5$
R Recycle ratio $= Q_R/Q_o$
Q_R Recycle flow rate

3. Facultative Pond [10, 11, 23]

A facultative pond is a type of oxidation pond which works under aerobic conditions in the upper layers while anaerobic conditions are maintained in the lower layers. In between the two, a facultative zone exists: this means that in the facultative zone, aerobic conditions exist during the day and anaerobic conditions exist at night. The oxygen for the upper surface comes from the re-aeration from air and from the photosynthesis of algae. The algae grow in the pond, since they utilize the nutrients from the waste and the light energy. The algae/oxygen is consumed by the bacteria in the pond for oxidation of waste. Thus, there is a symbiosis around the carbon dioxide, which is the end product of the bacterial action, and which is utilized by the algae. The BOD loading criteria for facultative ponds are given below:

Mass (lb/day) = Flow (MGD) × Concentration (mg/L) × 8.34 (lb/Mgal)/mg/L
BOD loading for the total system $\leq$ 35 pounds BOD$_5$/acre/day
Minimum = 3 ponds
Depth D = 3 to 8 ft
Minimum time t = 90 to 120 days

Fig. 1.24 Aerated Lagoon

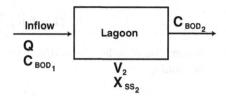

4. Aerated Lagoons (Fig. 1.24) [10, 11, 23]

When oxygen is supplied to ponds through a process of mechanical agitation, it is termed an aerated lagoon. Aerated lagoons can be described in simple terms as an activated sludge process unit in which there is no return pathway for the sludge.
Here

$$C_{BOD_2} = \frac{C_{BOD_1}}{1 + K_1 O} \qquad (1.98)$$

where

O $\dfrac{V_2}{Q}$

K_1 First-order rate constant for BOD removal, d^{-1}

1.4.4 Disinfection of Water [9]

The process of killing pathogens is called disinfection and is common to both water and wastewater treatment. The methods of disinfection are physical (light, heat) and chemical (ozone, chlorine). Chlorine can be applied either as a solution prepared from bleaching powder or as liquid gas under high pressure. The residual chlorine remaining after disinfection can be removed by sulfur dioxide or sodium bisulfite. Chlorine is frequently used in North America. Ozone is used commonly in Europe; it is more expensive but does not leave any residuals.

The time required to kill pathogens follows a first order law: the rate of time required to kill is proportional to the pathogens remaining:

$$-\frac{dN}{dt} = KN \qquad (1.99)$$

N: number of pathogens remaining; K is a constant

$$\text{Therefore, } \frac{dN}{N} = -K\,dt \tag{1.100}$$

If there are N pathogens at time $t = 0$, the pathogens at time t are given by integrating:

$$\ln\left(\frac{N_t}{N_0}\right) = -Kt \tag{1.101}$$

or

$$N_t = N_0 \exp(-Kt) \tag{1.102}$$

The disinfection efficiency can be estimated by:

$$C^n t = \text{a constant} \tag{1.103}$$

where
C Concentration of disinfectant
t Time required to kill a constant percentage of pathogens
n A coefficient of dilution

The effect of temperature on disinfection is according to the Arrhenius law:

$$\ln\left(\frac{t_1}{t_2}\right) = \frac{\Delta E(T_2 - T_1)}{RT_1 T_2} \tag{1.104}$$

where t_1 and t_2 are the times required to kill an equal percentage of pathogens at temperatures T_1 and T_2, respectively. R is the gas constant and $\Delta E = $ activation energy.

The concentration of organisms also affects the concentration of disinfectant needed:

$$C^q N_R = \text{a constant} \tag{1.105}$$

where
C Concentration of disinfectant
N_R Concentration of pathogens reduced by a fixed percentage at a given time t
q A coefficient

References

1. Henry, J. G., & Heinke, G. W. (1998). *Environmental science and engineering*. Upper Saddle River, New Jersey: Prentice Hall.
2. Davis, M. L., & Masten, S. J. (2004). *Principles of environmental engineering and science*. New York: McGraw Hill.

3. Gupta, R. S. (2001). *Hydrology and hydraulic systems* (2nd ed.). Prospect Heights: Waveland Press.
4. Viessman, W., & Hammer, M. J. (1985). *Water supply pollution control* (4th ed.). New York: Harper & Row.
5. Government Institutes. (2009). *Environmental law handbook* (19th ed.). Maryland: Littlefied Publishing.
6. McCabe, W. L., Smith, J. C., & Harriott, P. (2005). *Unit operations of chemical engineering* (7th ed.). New York: McGraw Hill.
7. Cussler, E. L. (2009). *Diffusion: Mass transfer in fluid operations* (3rd ed.). New York: Cambridge University Press.
8. Farrell, S., Hasketh, R. P., & Slater, C.S. (2003). Exploring the potential of electrodialysis. Chemical Engineering Education.
9. Reynolds, T. D., & Richards, P. A. (1996). *Unit operations and processes in environmental engineering* (2nd ed.). Boston, MA, PWS.
10. NCEES. (1998). *FE Reference Handbook* (8th ed). Seneca: National Council of Examiners for Engineering and Surveying.
11. Hammer, M. J., & Hammer, M. J., Jr. (2001). *Water and wastewater technology* (4th ed.). Upper Saddle River: Prentice Hall.
12. Olson, R. M., & Wright, S. J. (1990). *Essentials of fluid mechanics* (5th ed.). New York: Harper & Row.
13. Addison, H. (1964). *A treatise on applied hydraulics*. New York: John Wiley.
14. Comp, T. R. (1946). *Journal of Sewage Water*, p 183.
15. Masters, G. M. (1998). *Introduction to environmental engineering and science*. Upper Saddle River: Prentice-Hall.
16. Rose, H. E. (1950). On the permeability of cemented beds and sandstones. *Journal of Institution, Water Engineers, 4*(7), 535.
17. Rose, H. E., & Risk, A. M. A. (1949). Further researches in fluid flow through beds of granular materials. *Proceedings of the Institution of Mechanical Engineers, 160*, 493.
18. Rose, W. (1949). Theoretical generalizations leading to the evaluation of relative permeability. *Journal of Petroleum Technology*. Amer. Institute Min Engrs, Tech Paper 2563, p 111.
19. Treybal, R. E. (1980). Mass transfer operations (3rd ed.). New York: McGraw Hill.
20. Lin, W.-C., Rice, P. A., Cheng, Y.-S., & Barduhn, A. J. (1977). Vacuum Stripping of Refrigerants in Water Sprays. *AICHE Journal, 23*(4).
21. Naimpally, A. (1971). Vacuum stripping of butane from dilate butane-water solution (M.S. Thesis, Syracuse University, 1971).
22. Seader, J. D., & Henley, E. J. (1998). *Separation Process Principles*. New York: Wiley.
23. Tchobanoglous, G, et al. (2004). Metcalf & Eddy, Inc. Wastewater Engineering. New York: McGraw Hill.

Chapter 2
Groundwater, Soils, and Remediation

Groundwater is enclosed within the pores of soil, below the surface of the earth. The groundwater flows in what are called aquifers. The layer of soil above the aquifer could either be porous or be impermeable. In the former case, the surface of the aquifer would be at atmospheric pressure, and the aquifer is called an *unconfined aquifer*. Whenever the aquifer is overlain with impermeable soil, the pressure of water rises above atmospheric pressure; the aquifer is termed a *confined aquifer*.

The layer of soil above the water table is called the vadose zone.

Some relevant definitions are given below:

Aquitard. A water-saturated sediment or rock whose permeability is too low to transmit a useful amount of water.

Aquifuge. Impervious formation that neither contains water nor transmits it.

Aquiclude. A formation that is impermeable to water.

Unconfined aquifer. A groundwater formation with no overlying beds of low permeability that physically isolate it from the surface.

Confined aquifer is a groundwater formation which is overlain by impervious soil. The water within the confined aquifer is generally at a pressure greater than one atmosphere.

Groundwater table is the upper surface of the zone of saturation. In the zone of saturation, all the pores are completed filled with water.

Artesian aquifer is the confined aquifer whose pressure is high enough to push water to the surface when a well is dug.

Piezometric surface. The surface to which the water would rise or does rise in a well tapping an aquifer. For the case of unconfined aquifers, the piezometric surface will be slightly above the water table, due to the rise caused by surface tension (or capillary) forces. For confined aquifers, the piezometric surface is an imaginary surface to which the water layer would rise due to higher pressure within the aquifer. This can be computed using the principles of fluid statics.

Porosity is the ratio of the void volume (or pore volume) in a rock to the total volume:

A. Naimpally and K. S. Rosselot, *Environmental Engineering: Review for the Professional Engineering Examination*, DOI: 10.1007/978-0-387-49930-7_2, © Springer Science+Business Media New York 2013

$$\text{Porosity } \epsilon = \frac{V_v}{V}$$

V_v Void volume
V Total volume

Specific yield, S_y, of a rock is the ratio of the ultimate volume of water that can be released from the rock, to the total volume:

$$S_y = \frac{V_d}{V}$$

where
V_d Volume of drained water
V Total volume of rock

Specific retention, S_r, is the capacity of the rock to retain water:

$$S_r = \frac{V_r}{V}$$

where
V_r Volume of water retained by the rock
V Total volume of rock

Safe yield is the amount of water that can be drawn in one year without ultimately depleting the aquifer.

Gravity yield is the percentage of the total volume of rock occupied by groundwater that can drain under the action of gravity during a given time of drainage.

Permeability is a measure of the ease of flow of groundwater through aquifers. The standard coefficient of permeability (hydraulic conductivity), K, is the liters of water per day that will flow through one square meter cross section under a unit meter hydraulic gradient.

Coefficient of transmissivity, T, is the rate at which water will flow through a vertical strip of the aquifer one meter wide and extending through the full saturated thickness, under a hydraulic gradient of 1.00. The coefficient of transmissivity is a measure of how much water moves through the formation:

$$T = Kh$$

where
K Coefficient of permeability (hydraulic conductivity)
h Thickness of the aquifer

Coefficient of storage, S, is the volume of water released or taken into storage by the aquifer, per unit surface of the aquifer, per unit decline or rise of head. For confined aquifers, the volume of S is given by

$$S = \varepsilon\gamma h\left(\beta + \frac{\alpha}{\varepsilon}\right)$$

where

ε Porosity of aquifer
γ Unit weight of water
h Thickness of aquifer
β The compressibility or the reciprocal of the bulk modulus of elasticity of water
α Vertical compressibility of the material of the aquifer.

2.1 Soils

The porosity of some representative soils is given in the adjacent table, along with the specific yields of those soils. It may be noted that some soils like clay have a low specific yield even though they have a high porosity (Table 2.1).

2.2 Flow Fields

The velocity at any point in a flow field can be described in a more simple manner by using:

- The stream function.
- The velocity potential function.

Stream function $\chi(x, y)$ is used for describing the two-dimensional steady flow of incompressible fluids. The function $\chi(x, y)$ is given by:

Table 2.1 Porosity and specific yields of different materials

Materials	Porosity (%)	Specific yield (%)
Clay	45–55	3–4
Sand	30–40	25
Gravel	30	22
Gravel and sand	20–25	16
Limestone	1–10	2

Sources [2, 4, 5]

$$u(x,y) = x - \text{component of velocity} = \frac{\partial \chi}{\partial y}$$

$$v(x,y) = y - \text{component of velocity} = \frac{-\partial \chi}{\partial y}$$

Appropriate boundary conditions would need to be given for $u(x,y)$ and $v(x,y)$ in order to obtain the function $\chi(x,y)$.

Stream lines are lines of constant x values. A tangent to the stream line at any point for steady flow gives the velocity of the fluid at that point. A complete set of stream lines in a flow field is called a flow net.

Velocity potential, $\phi(x,y)$, is used to describe a two-dimensional, irrational flow of an incompressible fluid. The function $\phi(x,y)$ is defined by:

$$u(x,y) = \frac{\partial \phi}{\partial x} = x - \text{component of velocity}$$

$$v(x,y) = \frac{\partial \phi}{\partial y} = y - \text{component of velocity}$$

Darcy's Law is used to describe the flow as a function of the hydraulic gradient:

$$Q = -KA\left(\frac{dh}{dL}\right) \text{ for one dimensional flow.}$$

where
Q   Volumetric flow rate, m³/d
A   Cross-sectional area of flow, m²
$\frac{dh}{dL}$ Hydraulic gradient in the direction of flow
K   Coefficient of permeability (hydraulic conductivity) of the soil, m/d

Darcy velocity is the hypothetical velocity of the fluid, obtained by dividing the volumetric flow rate, Q, by the cross-sectional area of flow, A. It assumes that the water velocity is uniform across the cross-sectional area, A. (Fig. 2.1).

Fig. 2.1 Explanation of Darcy velocity

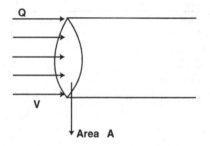

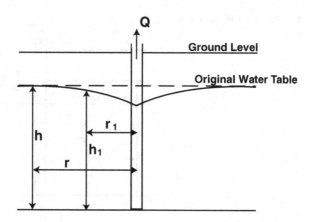

Fig. 2.2 Well in an unconfined aquifer

Actually, the area A consists of (solid) soil and the flowing water between the pores of the soil. The actual area used by the flowing groundwater is $A^1 = \epsilon_A$, where ϵ is the porosity of the soil.

Actual velocity, v^1, is the velocity obtained by dividing Q by A^1. Therefore,

$$\frac{Q}{A^1} = \frac{Q}{\epsilon_A} = \frac{1}{\epsilon}\left(\frac{Q}{A}\right) = \frac{v}{\epsilon}$$

Thus, the actual velocity, v^1, is the Darcy velocity V, divided by the porosity ϵ.

2.3 Wells in Aquifers

Case 1: Wells in unconfined aquifers (Fig. 2.2) [1–3].

A well in an unconfined aquifer having a water draw-out rate of Q $\left(\frac{\text{volume}}{\text{time}}\right)$ will cause a "cone of depression" in the surface of the groundwater table. For two radii, r and r_1, the heights of the groundwater table (modified, due to the presence of the well) can be stated as h and h_1, respectively (shown in Fig. 2.2). The formula for the cone of depression can be derived to be

$$Q = \frac{\pi K\left(h_1^2 - h^2\right)}{\ln\left(\frac{r_1}{r}\right)}$$

Case 2: Wells in confined aquifers (Fig. 2.3) (1,2).

For the case of confined aquifers, the thickness of the aquifer is assumed to be M, and the water table is a hypothetical (equivalent) one, since there is impervious

Fig. 2.3 Well in a confined
aquifer

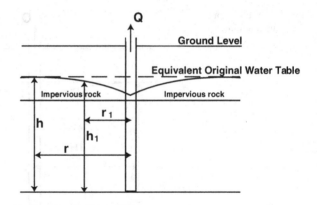

rock outside the confined aquifer. An artesian well used to draw out water at the
rate of Q will cause a change in the configuration of the equivalent water table
shown in the diagram.

Using the same nomenclature as before, the formula in this case is

$$Q = \frac{2\pi KM(h_1 - h)}{\ln\left(\frac{r_1}{r}\right)}$$

2.4 Remediation Technologies

2.4.1 Non-Aqueous Phase Liquids

Non-aqueous phase liquids (NAPLs) are organic materials, generally hydrocarbons, that are immiscible with water and do not dissolve in water. There is a large
variety of NAPLs: Light NAPLs (LNAPLs) move downward through the vadose
zone and float above the water table; dense NAPLs (DNAPLs) move downward to
the bottom of the water. NAPLs get trapped in individual pores of the soil–water
complex and thus are difficult to remove. It is difficult to detect the presence of
NAPLs at sites due to their being trapped in individual pores, and methods such as
monitoring wells and measurement of hydrocarbon in soil are not reliable by
themselves. Capillary forces make it difficult to remove NAPLs even when oil
field technology is utilized for this purpose. LNAPLs are generally from the
petroleum industry and involve low boiling, light hydrocarbons, like gasoline,
diesel, etc. DNAPLs generally originate from a variety of chemical manufacturing
processes, chlorinated solvents, etc.

2.4.2 Pump-and-Treat Technology

This method utilizes treatment processes being carried out above the ground on groundwater that has been pumped up for this purpose. Treated water can then be either used as such or pumped back to the aquifer. The pump-and-treat technology is useful for pollutants which have a low boiling point and thus can be removed easily with conventional technologies. It is ineffective with groundwater NAPLs.

Vacuum extraction is a technique used for the removal of low volatility organic compounds from soils. The first step in the process is to clearly define the field of ground that needs to be cleaned up. A number of extraction wells are then drilled into the contaminated field. Negative pressure (i.e., vacuum) is then applied to each extraction well. The range for each well generally extends from 100 to 150 feet. The volatile compounds from the soil are volatilized and extracted at the well. Water from groundwater below the well could also be extracted causing a mixture of air, water, and organic compounds to be removed at the well. In some cases, air may be pumped and fed into the soil in a programmed manner in order to help to cause steady air flow within the soil and induce a greater removal of volatile compounds from the soil. The time for the removal of volatile compounds can be computed using permeability data, gas flow assuming an ideal, potential flow, and the appropriate boundary conditions. The system variables that affect the removal rate are condition of the soil and the extent of contamination; type of soil and soil permeability, porosity, and other properties; types of contaminants and their properties; number of wells and extent of air pumping (sparging). The mixture of air, water, and contaminants removed at each extraction well is treated by several processes: separation of air and water; water treatment for the water; some type of oxidation of the volatile components, and the removal of the oxidized products before the air is vented.

Bioremediation processes are newer processes that are carried out in situ in the soil. Two conditions are needed if bioremediation is to be successful: The soil must have a large hydraulic conductivity and the microorganisms that are used must be able to degenerate the organic components present in the soil. Oxygen in the form of either air or oxygen needs to be injected into the soil in order for microorganisms to act upon the contaminants.

References

1. Bedient, PR., et al. (1999). *Groundwater contamination* (2nd Edn). Upper Saddle River: Prentice Hall.
2. Masters, G. M. (1998). *Introduction to environmental engineering and science* (2nd Edn). Upper Saddle River: Prentice-Hall.
3. NCEES. *FE reference handbook* (8th Edn). Clemson: National council of examiners for engineering and surveying, (2005).
4. Linsley, R. K., et al. (1992). *Water resources engineering* (4th ed.). NY: McGraw Hill.
5. Walton, W. C. (1970). *Groundwater resources evaluation*. NY: McGraw Hill.

Chapter 3
Air

3.1 Categories of Air Pollution

Air pollutants can be divided into two broad categories: particulate matter and gases. Particulates include small solid and liquid particles such as smoke, dust, and haze. Gases include carbon monoxide and VOCs. Pollutants can also be divided into primary pollutants and secondary pollutants. Primary pollutants are emitted directly from a source, while secondary pollutants are created in the atmosphere from primary pollutants. An example of a secondary pollutant is ozone, which is harmful to plant life and to lung tissue. Ozone forms in the ambient air from primary pollutants such as VOCs in the presence of sunlight.

Pollutants enter the air from both anthropogenic (human-caused) and non-anthropogenic sources. Volcanoes are an example of a non-anthropogenic source of air pollution.

Air pollution is also categorized by the type of source: stationary and mobile. Stationary sources include electric utilities and chemical manufacturing plants. Mobile sources include automobiles, airplanes, and boats. Mobile sources emit more air pollution than stationary sources in the United States.

Stationary sources usually emit pollutants from more than one area within the source. There may be several stacks along with evaporative ponds and building vents. Emissions from stationary sources are categorized as either fugitive or point emissions. Emissions from sources other than stacks, vents, and other point sources are known as fugitive emissions. One type of fugitive emissions is unintentional leaks from pipes, valves, flanges, and other process equipment. If this equipment is in good working order, these leaks are very small. However, because there can be thousands of such pieces of equipment in a plant, total fugitive emissions can be substantial.

Secondary emissions are emissions from stationary sources that are due to auxiliary activities at the facility. Pollutant emissions from wastewater treatment units at a refinery, for example, are called secondary emissions.

An area source is one that is too small to be considered on its own. Dry cleaners are an example of an area source.

A. Naimpally and K. S. Rosselot, *Environmental Engineering: Review for the Professional Engineering Examination*, DOI: 10.1007/978-0-387-49930-7_3, © Springer Science+Business Media New York 2013

3.2 Harmful Effects of Air Pollution

Air pollution can have harmful health and environmental effects. Health effects of air pollution include damage to the lungs, heart, circulatory system, and brain of affected individuals. The elderly, infants, pregnant women, and people with diseases of the heart and lungs are more susceptible to harm from air pollution than are other individuals. Children are particularly susceptible because their lungs are developing and because they spend more active time outdoors.

Acute health effects of air pollution, which are an immediate reaction to exposure and can be reversible, include nausea, headaches, and eye irritation. The effects of chronic exposure to air pollutants can include cancer, decreased lung capacity, and the development of asthma.

Air pollution also causes detrimental environmental effects that are an indirect threat to human welfare. Acid deposition occurs when sulfur dioxide and nitrogen oxides combine with water and fall to the earth. Acidification of water bodies can make them unsuitable for fish and can accelerate the leaching of harmful metals into water that is intended for human consumption. Acid deposition also harms trees and can cause decay of man-made structures.

High ozone levels harm lung tissue and are harmful to crops. Ozone is also particularly destructive to rubber and some fabrics.

More information on assessing the health effects of airborne contaminants is presented in Sect. 6.8.

3.3 Codes, Standards, Regulations, and Guidelines

There are six categories of pollutants that are defined as criteria air pollutants under the Clean Air Act and its amendments. These are particulate matter less than 10 μm in diameter (PM10), sulfur dioxide (SO_2), nitrogen oxides (NO_x), carbon monoxide (CO), ozone (O_3), and lead (Pb). National Ambient Air Quality Standards (NAAQS) must be met for these pollutants. These standards are based on concentration and time; higher concentrations are allowed for shorter periods of time. Table 3.1 gives the standards for criteria air pollutants.

The states are responsible for ensuring that the standards are met and must submit state implementation plans (SIPs) to the EPA. If an area does not meet an air quality standard, the state's plan for meeting the standard must be approved. Because criteria pollutants are emitted by stationary and mobile sources, plans for attaining the standards affect both types of sources.

Ozone is not emitted from any anthropogenic processes. Instead, it is created from nitrogen oxides and volatile organic compounds (VOCs) during the smog cycle. Because of their role in smog formation, VOCs are sometimes regulated under criteria air pollutant provisions of the Clean Air Act even though they are not a criteria air pollutant.

Table 3.1 Primary and secondary standards for criteria air pollutants

Pollutant	Primary standard		Secondary standard	
	Type of average	Standard-level concentration	Type of average	Standard level concentration
PM_{10}	24-h average not to be exceeded more than once per year on average over 3 years	150 μg/m³		Same as primary standard
$PM_{2.5}$	Annual mean, averaged over 3 years	12 μg/m³	Annual mean, averaged over 3 years	15 μg/m³
	98th percentile of the 24-h average concentration averaged over 3 years	35 μg/m³		Same as primary standard
Ozone	Annual fourth-highest daily maximum 8-h concentration, averaged over 3 years	0.075 ppm		Same as primary standard
Nitrogen dioxide	98th percentile of the 1-h average concentration averaged over 3 years	100 ppb		
	Annual mean	53 ppb		Same as primary standard
Sulfur dioxide	99th percentile of the 1-h daily maximum concentrations averaged over 3 years	75 ppb	3-h average concentration not to be exceeded more than once per year	0.5 ppm
Carbon monoxide	8-h average not to be exceeded more than once per year	9 ppm		No secondary standard
	1-h average not to be exceeded more than once per year	35 ppm		No secondary standard
Lead	Rolling 3 month average not to be exceeded	0.15 μg/m³		Same as primary standard

The new source review (NSR) program for criteria air pollutants requires all new major sources in and out of attainment areas to obtain construction permits. Major sources in non-attainment areas must apply control technologies to achieve the lowest achievable emission rate (LAER), regardless of cost. The best achievable control strategy (BACT) for a source, required in attainment areas, takes available technologies, energy and environmental impacts, and economics into account. BACTs comply with prevention of significant deterioration (PSD) regulations.

National emission standards for hazardous air pollutants (NESHAPs) also exist. Hundreds of compounds have been designated as hazardous air pollutants (HAPs). These compounds can cause irreversible or incapacitating effects at low concentrations. Maximum achievable control technologies (MACTs) and generally achievable control technologies (GACTs), which are less stringent than MACTs and are allowed for some area sources, apply to HAP. MACTs take economic considerations into account and are different for new and existing sources.

New Source Performance Standards (NSPSs) set emission limitations on new or substantially modified sources in certain industry categories.

Regulations can require the use of certain control equipment or process equipment, or they can mandate operating and maintenance procedures.

3.4 Meteorology

Wind and other meteorological phenomenon affect the way air pollutants are dispersed as well as the formation of some harmful air pollutants.

The earth's atmosphere is divided into four layers. The layer closest to the earth is called the troposphere. This layer is the thinnest of the four layers, but it contains most of the mass of the atmosphere and nearly all of the water in the atmosphere. The troposphere is where weather occurs and it is the most unsettled of the atmospheric layers. The depth of the troposphere is constantly changing. On average, it is 16.5 km deep at the equator and 8.5 km deep at the poles.

The stratosphere is immediately above the troposphere. The concentration of ozone (O_3) in the stratosphere is elevated. This "layer" of ozone absorbs ultraviolet light that causes skin cancer, cataracts, crop damage, and harm to some forms of aquatic life. Some halogenated organic compounds, such as CFCs, are very stable in the troposphere. This stability permits them to reach the stratosphere where they dissociate. The halogens they contain can then catalyze the destruction of ozone molecules, causing depletion of the ozone layer and a deterioration of its protective effects.

Insolation is the amount of incoming solar radiation that is received at a particular time and place. Insolation depends on four factors: (1) the solar constant, (2) the transparency of the atmosphere, (3) the duration of sunlight, and (4) the angle at which the sun's rays are striking the earth. A flat surface perpendicular to the sun's rays receives the most insolation.

The solar constant is the amount of radiation received at a point outside the atmosphere at the earth's mean distance from the sun and is 4,871 kJ/m²-h.

Transparency is determined by the extent to which the sun's rays are reflected back into space by clouds and the earth's surface, as well as by absorption of radiated solar energy by molecules and clouds in the atmosphere. Transparency varies with cloudiness, latitude, and season. On average, 30 % of the radiation from the sun is reflected back into space. Also on average, 3 % of the energy from the sun's rays is absorbed by clouds and another 16 % is absorbed by water vapor, dust, and ozone in the atmosphere.

Daylight duration also varies with latitude and with the season. At the equator, day and night are always of equal length. At the North Pole, a summer day lasts 24 h, and a winter night lasts 24 h.

The angle of the sun's rays affects insolation because when the rays strike the earth at an angle, the energy they possess is spread over a greater surface area. Time of day, the season, and latitude all affect the angle of the sun's rays.

The earth is constantly receiving energy from the sun and radiating energy out into space. If the earth receives more energy than it radiates, temperatures rise. While molecules and clouds in the atmosphere absorb some of the energy from the sun, they absorb even more energy that is being radiated from the earth. They then release this energy back to the earth or out into space. If it were not for this "greenhouse effect," the earth would be much colder. The most important absorbers of terrestrial radiation are clouds and water vapor.

The earth naturally goes through cycles of warmer temperatures and cooler temperatures. However, anthropogenic releases of molecules that are particularly effective at absorbing terrestrial radiation (greenhouse gases) may be causing the earth to become warmer at an unnaturally fast rate. Greenhouse gases include carbon dioxide, methane, nitrous oxide, chlorofluorocarbons (CFCs), and ozone.

Low-pressure areas are created because of greater insolation in some areas and because different features on the earth absorb heat from the sun differently. Wind is the result of air moving from areas of high pressure to low pressure. The rotation of the earth modifies this movement of air: Winds from the south in the northern hemisphere are deflected east, and winds from the north in the northern hemisphere are deflected west.

Wind at the earth's surface is slowed by friction, so that wind speed increases with altitude above ground level. The first 500–1,000 m of atmosphere above the earth's surface is called the atmospheric boundary layer. Within this layer, friction affects wind. Wind at the surface of the earth is slower than wind above the boundary layer. The smoother the surface of the earth is, the closer to the surface the maximum wind velocity is reached.

Winds are named according to the direction they come from, so that a north wind is a wind that blows from the north. The direction that most winds come from is called the prevailing wind.

Weather maps show isobars (lines where the pressure is the same). Around a low- or high-pressure area, the isobars form circles. If the pressure distance between lines is constant, wind speed is highest where the lines are close together. Wind does not blow perpendicular to isobar lines except at the equator because of the influence of the rotation of the earth.

Because friction effects decrease with increasing height above the earth's surface, wind at the atmospheric boundary layer blows in a slightly different direction than wind at the surface of the earth. The change in wind direction with altitude is called the Ekman spiral. Surface winds in the northern hemisphere blow toward low-pressure areas in a counterclockwise direction and away from high-pressure areas in a clockwise direction.

A cold front is the boundary between warm air and cool air when a cooler air mass is moving into an area, and a warm front is the boundary between warm air and cool air when a warmer air mass is moving into an area. Both types of fronts cause temperature inversions, where warm air is pushed over cooler air, trapping the cooler air beneath it. Temperature inversions are also created in valleys when cool air sinks to the bottom of the valley at night, trapping warm air above it. Inversions are significant to air pollution because they reduce vertical mixing in the atmosphere and allow pollutant concentrations to build up.

Building materials such as concrete and brick absorb heat from the sun more strongly than most natural features. Large cities experience a heat-island effect, where the temperature is always elevated. The city emits heat all night and begins to warm from the sun the following morning before it ever has a chance to cool down completely.

The lapse rate is the rate of change in temperature with height above the surface of the earth. In the troposphere, the average lapse rate is –6 to –7 °C/km. This means that the temperature drops 6 or 7 °C with each kilometer above the surface.

A stable atmosphere is one that resists vertical motion. Stable conditions occur at night when there is little or no wind. Unstable conditions occur on high insolation days with low wind speed, or in a low-pressure system.

Barometric pressure drops with an increase in elevation. At a distance above sea level of z, the pressure is

$$p_z = p_0 e^{-(8.4 \text{km})z}$$

In this equation, p_0 is the pressure at sea level. Because gases at atmospheric pressure obey the ideal gas law, the partial pressures of the gases that make up the atmosphere drop with increasing elevation at the same rate as the overall atmosphere. The effect of elevation on the molar concentration of a compound in air can be written as

$$X_z = X_0 e^{-(8.4 \text{km})z}$$

3.5 Atmospheric Chemistry

Dry air consists of about 78 % nitrogen and 21 % oxygen. Trace amounts of carbon dioxide, methane, argon, helium, and hydrogen make up most of the remaining 1 %.

At steady state, the flux of material into the atmosphere is the same as the flux of a material out of the atmosphere. The steady-state atmospheric residence time is the total amount of material in the atmosphere divided by the steady-state flux of material either into the atmosphere or out of the atmosphere, or

$$\tau = \frac{M}{F}$$

The formation of photochemical smog in the troposphere is of great concern because photochemical smog contains a high concentration of oxidizing agents. One of these oxidizing agents is ozone, or O_3, which is short-lived in the lower atmosphere. Photochemical smog can cause eye irritation, poor visibility, and breathing difficulties.

The formation of photochemical smog requires ultraviolet light and sufficient concentrations of hydrocarbons and nitrogen oxides. Hydrocarbons and nitrogen oxides are emitted during the burning of fossil fuels.

Many chemical reactions are involved in photochemical smog production. These reactions are referred to as the "smog cycle" because some of the reactants are recreated to participate again in the cycle. Some of the reactions are listed here:

$$HO^{\cdot} + RH \rightarrow H_2O + R^{\cdot}$$
$$R^{\cdot} + O_2 \rightarrow RO_2^{\cdot}$$
$$HO_2^{\cdot} + NO \rightarrow NO_2 + HO^{\cdot}$$
$$RO_2^{\cdot} + NO \rightarrow NO_2 + RO^{\cdot}$$
$$O_3 + NO \rightarrow NO_2 + O_2$$
$$O_3 + hv \rightarrow O^* + O_2$$
$$NO_2 + hv \rightarrow NO + O$$
$$O^* + H_2O \rightarrow 2OH^{\cdot}$$
$$CO + OH^{\cdot} + O_2 \rightarrow CO_2 + HO_2^{\cdot}$$

In these equations, R is any hydrocarbon group, such as a methyl group, and hv is light energy. The symbol "·" indicates an unpaired electron. The net result of these chemical reactions is elevated levels of ozone and nitrogen dioxide.

Hydroxyl radicals, represented by $HO\cdot$, have the potential to oxidize most organic compounds in the presence of light. Oxidation by hydroxyls is a major removal mechanism for most organic compounds in the atmosphere. In fact, a compound's half-life in the atmosphere is generally estimated by combining its reaction rate with hydroxyl radicals and the concentration of hydroxyl radicals in the atmosphere, as follows.

$$t_{1/2} = \frac{0.693}{k_{OH}[\cdot OH]}$$

The average concentration of hydroxyl radicals in the atmosphere is generally taken to be 8×10^5 molecules/cm^3.

The reaction for formation of hydroxyl radicals is

$$O_3 + hv \rightarrow O^* + O_2$$
$$O^* + H_2O \rightarrow 2HO\cdot$$

In these equations, O* indicates photochemically excited oxygen atoms. Compounds that resist oxidation by hydroxyl radicals, such as CFCs, are likely to persist in the atmosphere for a long time. Besides reaction with hydroxyl radicals, compounds can be removed from the atmosphere through physical means (wet or dry deposition), by reaction with nitrate or ozone, and by photolysis.

Industrial smog, or gray smog, is a mixture of sulfur oxides, particulates, and fog that results in the formation of sulfuric acid droplets and particulates coated with sulfuric acid. This respirable sulfuric acid is extremely harmful to lung tissue. In 1952, London had an episode of this type of smog that resulted in 4,000 deaths.

The worst smog episodes tend to occur on sunny days in cities that are surrounded by mountains during what is called a temperature inversion. During a temperature inversion, the air above the city is not allowed to mix vertically. If mountains surround the city, horizontal mixing of the air is also confined and concentrations of pollutants can become much higher than normal.

The catalyzed destruction of ozone molecules, which is occurring in the stratosphere due to the introduction of stable halogenated compounds such as CFCs, is illustrated in these equations.

$$Cl + O_3 \rightarrow ClO + O_2$$
$$ClO + O \rightarrow Cl + O_2$$

Nitric acid is formed by the reaction of nitrogen dioxide with hydroxide radicals, as follows.

$$\cdot OH + NO_2 \rightarrow HNO_3$$

Sulfuric acid is formed when sulfur dioxide is oxidized by oxygen or hydrogen peroxide.

3.6 Measuring Air Pollution

The concentration of dilute materials in air is often given in parts per million (ppm), parts per billion (ppb), or parts per trillion (ppt). For gases, these "parts" are on a volume basis, and parts per million might be written as ppmv. (This is in contrast to the meaning of ppm for solids and liquids, which are generally given on a mass basis.) For ideal gases, a volume basis and a molar basis are equivalent, because for all ideal gases at the same temperature and pressure, volume per mol is a constant.

Converting from ppmv to molar concentrations is done using molar volume at standard temperature and pressure (22.4 L) and then adjusting for non-standard temperature and pressure effects using the ideal gas law.

Converting from ppmv to mass concentrations is done by multiplying ppmv by the molecular weight of the compound and dividing by the molar volume of air.

Emissions of pollutants must often be quantified in order to comply with air quality regulations. Sometimes regulations require that the emissions be measured or that ambient air concentrations be monitored. Measurement and identification of chemical species in gaseous streams exiting from stacks can be expensive and problematic. Measuring emissions from sources other than stacks can be even more difficult, or even impossible.

The EPA has developed Federal Reference Methods and Federal Equivalent Methods for sampling and analysis of pollutants in the ambient air and from emission sources. The methods specify precise procedures that must be followed for any monitoring activity related to the compliance provisions of the Clean Air Act. The procedures regulate sampling, analysis, calibration of instruments, and calculation of emissions.

One of the most common instruments used to detect and measure the presence of compounds present at low concentrations is a chromatograph. Chromatography can be performed on gas or liquid samples. The sample first passes though a column designed to separate the compounds in the sample. In general, smaller compounds pass through the column faster than larger compounds. Some compounds of similar size will pass through the column at the same rate.

As the sample passes through the column, a method of quantifying the amount of material leaving the column as a function of time is applied. Usually, the material is ionized as it leaves the column. Different methods of ionization are used for different samples, depending on what kinds of compounds are in the sample. The resulting signal is proportional to the quantity of compound present.

The result of chromatography is a chart that shows a series of peaks and the elution time for each peak. Each peak represents a compound (or a group of compounds if there is a similar-sized compound in the sample that passes through the column at the same rate). The peaks have different heights or areas, and these heights or areas are proportional to the concentration of compound present in the sample.

In spectrophotometry, a beam of light passes through a sample and the wavelength and amount of light absorbed are measured to determine the amount of a compound that is present. Sulfur dioxide concentrations are measured using spectrophotometry. A variation in spectrophotometry, called monochromatic absorption spectrophotometry, uses ultraviolet light with a wavelength of 253.7 nm to determine the amount of ozone in a sample.

In atomic absorption spectrometry (AAS), a sample is dissolved and atomized and then passed through a beam of light whose wavelength is specific to the chemical being tested for. The amount of light absorbed by the atomized matter and the wavelength emitted by the sample are used to identify the elements that are present and determine their concentration in the sample. AAS can be used to analyze for over 60 metals and metalloid elements.

Chemiluminescence measures the spectrum of light emitted by species that are undergoing specific chemical reactions. Nitrogen oxide and ozone form an excited form of nitrogen dioxide that emits a particular wavelength of light. The amount of light emitted and the total amount of oxygen present together are used to determine the amount of NO and NO_2 in a sample.

Fourier transform infrared spectroscopy (FTIR) can be used to directly measure over 100 species in ambient air. Every gas has a unique infrared absorption spectrum, and this technique can analyze for these spectra with great sensitivity.

Gravimetric analysis of particulate releases consists of weighing filters before and after the collection of particulates. The particulates can then be analyzed for their chemical makeup using many different techniques.

Cascade impactors are used to determine particle size distributions. Optical particle counters provide information on both the number of particles in a sample and their size. Optical particle counters work by illuminating particles with laser light. The scattering of the light is detected, and its pulse height is correlated with particle size. The flow rate of the sample is combined with information on the number of particles detected in order to determine particle concentration. As with many particulate measuring instruments, it is important to minimize bends in the sample stream. Such bends favor the removal of particles of large size and result in incorrect particle size distributions. Because optical particle counters register one particle at a time, a high concentration of particles in the sample stream will overload the instrument. Highly concentrated samples must be diluted to an acceptable level.

Methods for estimating emissions without using monitoring data or using limited monitoring data have been devised. One means of estimating emissions is to apply emission factors. For example, the equation for estimating fugitive emissions from organic chemical manufacturing plants using emission factors is

$$E_{TOC} = F_A WF_{TOC} N$$

In this equation, E_{TOC} is the emission rate of total organic compounds, F_A is the emission factor (which is specific to the type of equipment, given in kg/h/source), WF_{TOC} is the average weight fraction of total organic compounds in the streams serviced by the equipment, and N is the number of pieces of equipment.

Many emission factors depend on equipment throughput and are given in terms of mass of emission per volumetric flow rate of material through the equipment. Emission factors are usually available for equipment that has no emission controls. These factors were intended to be conservative and to apply to equipment that has no emission controls. As a result, emission factors generally result in an estimate of emissions that is higher than the actual emissions. Sometimes, factors are provided to adjust for techniques that reduce emissions.

3.7 Pollutant Dispersion

Many times, it is of practical interest what the concentration of a pollutant emitted at one point will be downwind of its source. When pollutants leave a source, they spread out and are carried by the wind, so that the pollutant stream becomes less concentrated but occupies a greater volume. The space that emissions from a source occupy is called a plume.

Table 3.2 Atmospheric stability under various conditions (US EPA [3])

Surface wind speed[a], m/s	Day solar insolation			Night Cloudiness[e]	
	Strong[b]	Moderate[c]	Slight[d]	Cloudy (≥4/8)	Clear (≤3/8)
<2	A	A – B[f]	B	E	F
2–3	A – B	B	C	E	F
3–5	B	B – C	C	D	E
5–6	C	C – D	D	D	D
>6	C	D	D	D	D

Notes [a] Surface wind speed is measured at 10 m above the ground
[b] Corresponds to clear summer day with sun higher than 60° above the horizon
[c] Corresponds to clear summer day with a few broken clouds or a clear day with sun 35–60° above the horizon
[d] Corresponds to a fall afternoon, or a cloudy summer day, or clear summer day with the sun 15–35°
[e] Cloudiness is defined as the fraction of sky covered by the clouds
[f] For A – B, B – C, or C – D conditions, average the values obtained for each
A = very unstable
B = moderately unstable
C = slightly unstable
D = neutral
E = slightly stable
F = stable
Regardless of wind speed, Class D should be assumed for overcast conditions, day or night

The speed at which pollutants from a source disperse depends in part on atmospheric stability. As discussed earlier, atmospheric stability is a function of wind speed and of how much energy the sun is providing. Table 3.2 lists atmospheric stability classes for varying conditions. This table shows that a still day with intense sunlight is classified as very unstable and is assigned an atmospheric stability of Class A. Class A atmospheric stability leads to plumes that disperse the most. These plumes start to spread out as soon as they are emitted and pollutant concentrations downwind of the source drop sharply. During highly unstable conditions, a pollutant plume may rise and fall as it travels downwind. This is called a looping plume and can create unexpectedly high concentrations of pollutants at ground level. In contrast, on a still, clear night, which is classified as stable and assigned an atmospheric stability of Class F, plumes do not appreciably narrow after they are emitted, and the concentration of pollutants within the plume is maintained downwind of the source. A pollutant plume emitted during stable conditions is called a fanning plume. Because it occupies a narrow vertical band, it may travel far downwind without a great deal of dispersion. A coning plume is one that develops during neutral or slightly stable conditions. It spreads both up and down after being emitted from the source.

If a pollutant plume is released above an inversion boundary, it will disperse upward but not toward the ground. This is called a lofting plume. However, if a pollutant plume is released below an inversion boundary, it will disperse toward the ground but not upward beyond the inversion boundary. This is called fumigation.

Fig. 3.1 Vertical standard
deviations of a plume (US
EPA [3])

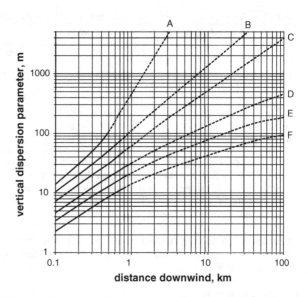

In practice, complicated computer models are used to predict concentrations of pollutants emitted from a source. A Gaussian model is a simplified model for estimating atmospheric dispersion of pollutants from a source. The Gaussian model for concentration estimates the concentration of pollutants in three dimensions: the distance downwind of the source along the plume's centerline, x, the horizontal distance from the plume centerline, y, and the vertical distance from ground level, z. The equations presented here do not take removal processes into account.

Vertical and horizontal standard deviations of a plume are needed when performing Gaussian modeling and are generally found in charts that provide the value of the standard deviation as a function of the distance downwind of the source. Curves for each atmospheric stability class are provided on these charts. Greater atmospheric stabilities result in smaller plumes and lower the standard deviations. Figures 3.1 and 3.2 give vertical and horizontal standard deviations of a plume.

Dispersion is also affected by how high above the ground the pollutant is emitted. Many times, air pollutants are emitted from stacks. The buoyancy the plumes possess at the point of release (often due their elevated temperature) carries them above the top of the stack. The effective stack height of a plume is the physical stack height plus the plume rise, or

$$H = h + \Delta h$$

The plume rise can be approximated as

$$\Delta h = \frac{1.6 F^{1/3} x^{2/3}}{\mu}$$

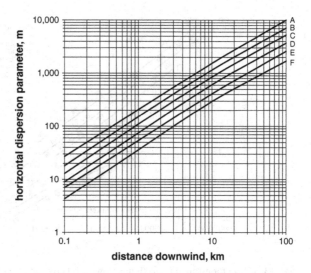

Fig. 3.2 Horizontal standard deviations of a plume (US EPA [3])

In this equation, F is the buoyancy flux.

The formula for calculating concentration of pollutants using Gaussian atmospheric dispersion modeling is

$$C = \frac{Q}{2\pi\mu\sigma_y\sigma_z}\exp\left(-\frac{1}{2}\frac{y^2}{\sigma_y^2}\right)$$
$$\times \left(\exp\left(-\frac{1}{2}\frac{(z-H)^2}{\sigma_z^2}\right) + \exp\left(-\frac{1}{2}\frac{(z+H)^2}{\sigma_z^2}\right)\right)$$

The downwind distance along the plume centerline, x, does not appear explicitly in this equation. This variable is accounted for because the values of the standard deviations, σ_y and σ_z, increase with increasing x. In this equation, μ is the average wind speed at the height of the stack and Q is the mass flow rate of the pollutant.

To find the concentration of a pollutant along the plume's centerline at ground level, y and z are set equal to zero, and the equation simplifies to

$$C = \frac{Q}{\pi\mu\sigma_y\sigma_z}\exp\left(-\frac{H^2}{2\sigma_z^2}\right)$$

The maximum concentration at ground level occurs when σ_z is equal to the effective stack height divided by the square root of two, or

$$\sigma_z = \frac{H}{\sqrt{2}}$$

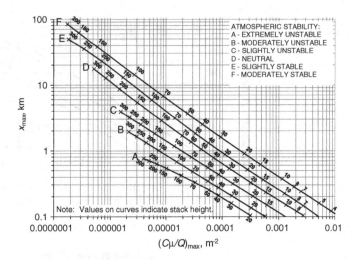

Fig. 3.3 Curve for estimating maximum downwind concentration from a plume; numeric values adjacent to curves are effective stack height (US EPA [3])

There are also charts, such as the one pictured in Fig. 3.3, for finding the maximum ground-level concentration of a pollutant as function of stack height and atmospheric stability class. The distance from the stack at which the maximum ground-level concentration exists can also be read from these charts. The maximum ground-level concentration can also be estimated using this equation:

$$\left(\frac{C\mu}{Q}\right)_{\max} = \exp\left(a + b \ln H + c(\ln H)^2 + d(\ln H)^3\right)$$

The curve-fitting constants a, b, c, and d in this equation, which are specific to each atmospheric stability class, are given in Table 3.3.

Table 3.3 Values of curve-fit constants for estimating $(C\mu/Q)_{\max}$ from H (in m) as a function of atmospheric stability

Stability	Constants			
	a	b	c	d
A	−1.0563	−2.7153	0.1261	0
B	−1.8060	−2.1912	0.0389	0
C	−1.9748	−1.9980	0	0
D	−2.5302	−1.5610	−0.0934	0
E	−1.4496	−2.5910	0.2181	−0.0343
F	−1.0488	−3.2252	0.4977	−0.0765

3.8   Risk Management Programs

The Clean Air Act as amended requires many facilities to determine the off-site consequences of an accidental release. Facilities that handle, manufacture, use, or store certain toxic or flammable substances above the specified threshold quantities are required to develop and implement a risk management program (RMP) that prevents the accidental release of theses substances and that mitigates the consequences of a potential release.

Off-site consequence analyses must be provided to state, local, and federal agencies for a worst-case release scenario and for alternate scenarios. A worst-case scenario is defined as the release of the largest quantity of a regulated substance from a single vessel or process line failure that results in the greatest distance to an endpoint. Examples of the distance to the endpoint include the distance that a cloud of toxic vapor, heat from a fire, or blast waves from an explosion will travel before dissipating to the point that serious injuries from short-term exposures will no longer occur. Meteorological and release conditions that result in the farthest endpoint are chosen when determining worst-case scenarios. For toxic gases, the worst-case release is one in which all of the gas is released at ground level in a 10-min period. For toxic liquids, the worst-case release is when all of the liquid in the process is spilled onto a flat surface where it spreads to a depth of one centimeter immediately. For flammable substances, the worst-case release is one that results in detonation of a vapor cloud that contains the entire amount of the flammable substance in the process and that is between the upper and lower flammability limits. Alternative release scenarios are scenarios that are more likely to occur than the worst-case scenario.

Generally, the risk management program rule allows for generic modeling of the potential fate of releases rather than site-specific modeling. Generic modeling is simpler and requires less data gathering, but it tends to overestimate the distance to endpoints.

The rule defines three program levels. Processes are generally eligible for Program 1 if there are no public receptors within the distance to the endpoint for the worst-case scenario. For processes subject to Program 2 or Program 3, both worst-case release scenarios and alternative release scenarios are required.

There are many means for reducing the potential for off-site consequences. Administrative controls that reduce the amount of a regulated substance stored on site or contained in a process reduce the potential for off-site consequences and may result in causing a facility to fall below the RMP reporting threshold. Passive mitigation systems are those that function without human, mechanical, or other energy input and include building enclosures or dikes and containment walls for reducing evaporation from liquids. These systems reduce the potential for off-site consequences. Modifications to materials that cause them to transport less effectively in the environment can also reduce the potential for off-site consequences. All of these measures might result in a reduction in a facility's program level.

3.9 Air Transport Systems

The formulas for volume that are presented in Table 3.4 may be useful in performing calculations.

The volumetric flow rate of gas in a duct is the average velocity multiplied by the cross-sectional area of the duct.

Standard temperature is 20 °C, and standard pressure is 1 atm. Pressure and temperature are evaluated at both absolute and relative scales. Monitoring data are frequently gathered at relative scales (such as gage pressure), but many engineering calculations require the use of absolute scales (such as absolute pressure).

Dry gas flow rates can be converted to wet gas flow rates if the moisture content expressed as a volumetric fraction of the stream is known, as follows.

$$\text{dry flow rate} = \text{wet flow rate}(1 - \text{moisture content})$$

Dry concentration can be calculated from wet concentration, as follows.

$$\text{wet concentration} = \text{dry concentration}(1 - \text{moisture content})$$

Again, in this equation, moisture content is expressed as a volumetric fraction of the stream.

Flow rates at standard temperature and pressure can be converted to actual flow rates by applying the ideal gas law.

$$\text{actual flow rate} = \text{standard flow rate}\left(\frac{T}{293\,\text{K}}\right)\left(\frac{1.013 \times 10^5\,\text{Pa}}{P}\right)$$

In this equation, T is actual temperature in degrees Kelvin and P is actual absolute pressure in Pa.

The viscosity of air can be estimated by using the following equation.

$$\mu = 51.12 + 0.372T + 1.05 \times 10^{-4}T^2$$
$$+ 53.147(\text{volumetric oxygen fraction}) - 74.143(\text{volumetric moisture fraction})$$

In this equation, T is actual temperature in degrees Kelvin.

Table 3.4 Formulas for calculating the volume of various shapes

Shape	Formula
Sphere	$\dfrac{4\pi r^3}{3}$
Cylinder	$H\pi r^2$
Pyramid	$\dfrac{LWH}{3}$
Cone	$\dfrac{H\pi r^2}{3}$

Just as with liquids, the Reynolds number is used to determine whether air flow is turbulent or laminar. The formula for the Reynolds number is

$$\text{Re} = \frac{Lv\rho}{\mu}$$

In this equation, L is the diameter, v is the velocity, μ is the viscosity, and ρ is the density. A Reynolds number higher than 10,000 is associated with turbulent flow.

Hoods are used to capture emissions and route them outside or to a control device. A hood's capture velocity is the air velocity at any point in front of the hood or at the hood opening that is necessary to overcome opposing air currents and to pull the contaminated air at that point into the hood. Velocity decreases rapidly with increasing distance from the hood. At one hood diameter away from the hood, gas velocities are roughly 10 % of what they are at the hood. Because of this, hoods must be very close to the emission source. Higher velocities are needed for large particles than for capturing individual compounds because individual compounds have little inertia.

For a freely supported hood without a flange, the hood capture velocity at distances nearer than 1.5 times the hood diameter can be estimated using the following equation.

$$Q = v_h\left(10x^2 + A\right)$$

In this equation, Q is the volumetric flow rate through the hood, v_h is the hood capture velocity at distance x from the hood, x is the distance to the farthest point of contaminant release, and A is the area of the hood opening.

For a hood with a wide flange, the equation for estimating hood capture velocity is

$$Q = 0.75\, v_h\left(10x^2 + A\right)$$

Flanges improve hood performance because they block crosscurrents and block clean air from entering the hood.

Hood static pressure, which is measured in the duct immediately downstream from the hood opening, is used to measure the velocity of air through the hood. Hood static pressure depends on the hood velocity and on the hood entry loss, which is dependent on the geometry of the hood. Assuming that slot loss is insignificant in comparison with duct entry loss (so that hood entry loss is equal to duct entry loss) yields the following equation for relating hood static pressure with hood velocity.

$$p_{s,\text{hood}} = -p_{v,\text{hood}} - F_{\text{duct}}\, p_{v,\text{hood}}$$

In this equation, $p_{s,\text{hood}}$ is the hood static pressure (the pressure being measured), $p_{v,\text{hood}}$ is the duct velocity pressure, and F_{duct} is the loss coefficient for the duct entry loss. The velocity of the air stream can be calculated from the duct velocity pressure and the density of the stream.

When transporting particulate matter through a duct, it is important to maintain a velocity that prevents the particulates from falling out of the gas stream and building up in the duct.

Fans have to be sized so that they provide hoods with the necessary transport velocity and maintain the necessary velocity through the duct. The rotational speed of a fan depends on the rotational speed of the motor as well as the diameter of the fan blades compared to the diameter of the motor's sheave, as follows.

$$n_{fan} = n_{motor} \frac{D_{motor}}{D_{fan}}$$

In this equation, n_{fan} is the rotational speed of the fan, n_{motor} is the rotational speed of the fan, and D is diameter. If no other parameters change, air flow rate and fan speed are directly related, as follows.

$$\frac{Q_2}{Q_1} = \frac{n_2}{n_1}$$

Q in this equation is volumetric flow rate.

The static pressure rise across a fan is given by

$$\Delta p_s = p_{s,outlet} - p_{s,inlet} - p_{v,inlet}$$

In this equation, p_s is static pressure and p_v is velocity pressure. If no other parameters change, the static pressure rise across a fan is directly related to the square of the fan's rotational speed, as follows.

$$\frac{\Delta p_{s1}}{\Delta p_{s2}} = \left(\frac{n_1}{n_2}\right)^2$$

Fan power is related to the cube of the fan rotational speed, as follows.

$$\frac{P_2}{P_1} = \left(\frac{n_2}{n_1}\right)^3$$

3.10 Treatment Technologies

Fugitive emissions can be reduced by regularly inspecting equipment and making needed repairs. The inspection is accomplished using meters that measure the concentration of organic compounds near the equipment. If the concentration is higher than a given amount, called the "leak definition," a repair is made.

Many different control technologies for removing pollutants from stack air releases exist. Acidic gases can be removed via wet scrubbing, which can be accomplished using either lime or caustic soda in solution. In lime scrubbing, lime (CaO) and water form calcium hydroxide which is then contacted with the air stream in a tray tower or packed tower. When sulfur dioxide is the acidic gas being removed by lime scrubbing, calcium sulfate is formed. Caustic scrubbing (using caustic soda or NaOH) is another means of removing acidic gases.

Dry scrubbing is also used to remove acidic gases from air releases. In dry scrubbing, calcium hydroxide slurry is injected into a chamber as tiny droplets that evaporate. (This slurry is more concentrated than the solution used in wet scrubbing.) As the pollutant stream passes through this chamber, the acid gases it contains are contacted with the calcium hydroxide droplets and are neutralized. They form particles that must be removed in a downstream step.

Choices for controlling particulate matter releases depend strongly on particle size. Health effects also depend strongly on particulate size. An aerosol is a particle that can remain suspended in air indefinitely.

If a particle is spherical, then its diameter characterizes its size. However, most particles are not spherical, and a uniform means of evaluating particle size is needed. A particle's Stokes diameter is the diameter of a spherical particle of the same density as the particle that has the same terminal settling velocity as the particle of interest. For spherical particles, the actual diameter and Stokes diameter are the same. Settling velocity depends on both particle size and particle density, which led to the development of a parameter called aerodynamic diameter. A particle's aerodynamic diameter is defined as the diameter of a spherical particle having a density of 1 gm/ cm^3 that has the same terminal settling velocity in the gas as the particle of interest. The aerodynamic diameter is related to the Stokes diameter, as follows.

$$D_a = D_s \sqrt{\rho}$$

In this equation, D_a is the aerodynamic diameter, D_s is the Stokes diameter, and ρ is the density of the particle.

A particle's terminal settling velocity is the velocity at which the particle approaches the ground in still air after gravitational pull is balanced by drag on the particle. For small particles (aerodynamic diameter less than 80 μm) in still air (Reynolds number less than 2), this velocity is given by the following equations.

$$v_t = \frac{g\rho D_s^2 C_c}{18\mu} = \frac{g D_a^2 C_c}{18\mu}$$

In these equations, g is the gravitational constant, D_a is the particle's aerodynamic diameter, D_s is the particle's Stokes diameter, C_c is the Cunningham slip correction factor, ρ is the particle's density, and μ is the viscosity of the gas. The Cunningham slip correction factor accounts for particles that are small enough to slip between gaps in the gas molecules and is significant for particles that have an actual diameter of 1 μm or less in air.

When discussing particle sizes, it is common to use aerodynamic diameter. PM_{10} and $PM_{2.5}$, for example, refer to particulate matter of aerodynamic diameter

Table 3.5 EPA particle size categories

Category	Aerodynamic diameter (µm)
Supercoarse	>10
Coarse	>2.5 and ≤10
Fine	>0.1 and ≤2.5
Superfine	≤0.1

10 µm and less and to particulate matter of 2.5 µm and less, respectively. Table 3.5 gives the EPA categorization for various particle categories. Note that in terms of particulate control, particles smaller than 10 µm are considered small. The smaller the particle is, the more difficult it is to remove.

Total suspended particulate matter (TSP) includes all particles up to 30 µm in diameter. PM_{10} particles are of particular concern because they enter the lungs, where they can become trapped. Particles larger than PM_{10} are generally filtered out before they reach the lungs, while particles smaller than 0.1 µm are generally expelled without being captured in the lungs. Fine and superfine particles can remain airborne for days and travel far from their area of origin. Particulate matter smaller than 0.1 µm is generated in many industrial processes, but particles this small tend to agglomerate so that very little particulate matter remains in this size range. Condensable particulate matter is emitted not as particles but as compounds that condense to form particulates after they are created. Most of the particulate matter formed from condensable particulate matter is in the troublesome $PM_{2.5}$ size range.

Particles emitted from anthropogenic sources often have a particle size distribution that fits a lognormal curve (they create a bell curve when they are plotted on a logarithmic scale).

Particles are created in a range of processes, and the nature of their origin can be helpful in determining ways to prevent their creation or remove them before discharge. During grinding processes, particles are released. These particles will be essentially identical in composition to the material being ground with some composed of the contact surface material. Most grinding particles are in the supercoarse range.

In an ideally functioning oil-fired boiler, ash and char in the 1–100 µm range are created and no organic vapor remains unburned. In actual practice, however, there is always some unburned fuel, which can then condense as the stream cools. This condensation is called nucleation and creates particles in the fine and superfine range that are very difficult to remove.

Sometimes contact between liquids and hot gas streams causes droplets to evaporate, releasing the solids they contain as particulate matter.

Cyclones such as the one pictured in Fig. 3.4 are a type of control device that are used to remove particles from gaseous streams. Cyclones are designed to cause a substantial drop in the velocity of the gas stream they are treating, which causes the particles to drop out of the gas stream. Because this separation process depends on the inertia of the particles, cyclones are not effective for small particles and in fact have little effect on particles smaller than 15 µm. Additionally, cyclones are not a good choice for treating a gas stream whose particles stick to the walls of the cyclone instead of dropping out of the gas stream.

Table 3.6 Cyclone ratio of dimensions to body diameter

Dimension	High efficiency	Conventional	High throughput
Inlet height H	0.44	0.50	0.80
Inlet width W	0.21	0.25	0.35
Body length L_b	1.40	1.75	1.70
Cone length L_c	2.50	2.00	2.00
Vortex finder length S	0.50	0.60	0.85
Gas exit diameter D_e	0.40	0.50	0.75
Dust outlet diameter D_d	0.40	0.40	0.40

Cyclone dimensions depend on whether the cyclone is high efficiency, conventional, or designed for high throughput. Table 3.6 provides typical cyclone dimensions expressed as a ratio of dimension to body diameter.

The efficiency of a cyclone depends on factors such as the viscosity and density of the stream, the dimensions of the cyclone, the inlet velocity, the density of the particles, and the number of effective turns the gas being treated makes in the cyclone. An equation for estimating the effective number of turns is

$$N_e = \frac{L_b + \frac{L_c}{2}}{H}$$

In this equation, L_b is the length of the cyclone's body, L_c is the length of the cyclone's cone, and H is the inlet height (circumference) of the cyclone.

A cyclone's particle cut size, or d_{pc}, is the diameter of particles that are collected with 50 % efficiency in that cyclone. The particle cut size can be predicted using the following equation:

$$d_{pc} = \left(\frac{9\mu W}{2\pi N_e v_i \left(\rho_p - \rho_g \right)} \right)^{0.5}$$

In this equation, μ is the viscosity of the gas, W is the inlet width of the cyclone, v_i is the inlet velocity of the gas entering the cyclone, and ρ_p and ρ_g are the density of the particle and the gas, respectively.

Once a cyclone's particle cut size has been calculated, the cyclone's collection efficiency for a particle of diameter d_p can be read from the cyclone efficiency curve, given in Fig. 3.5 or calculated using this equation:

$$\eta = \frac{1}{1 + \left(\frac{d_{pc}}{d_p} \right)^2}$$

Fabric filters, or baghouses, are sometimes used by themselves or are used downstream of cyclones for removing particles too small to have been removed by

Fig. 3.4 A cyclone

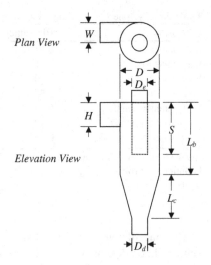

Plan View

Elevation View

Fig. 3.5 The cyclone
efficiency curve

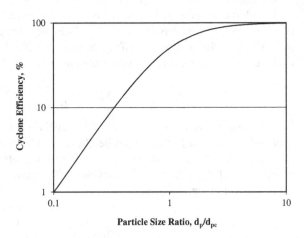

the cyclone. The filters trap particles but allow gases to pass through. Baghouse materials have been developed that can remove particles in the submicron range.

The dust being removed by a baghouse filter would accumulate and prevent even gases from passing through the filtration medium if this dust were not removed. Shaker mechanisms, which physically shake the bag in order to cause the dust it has trapped to fall down and be removed, are one method of dust removal. Sometimes a pulse of compressed air is used to cause the dust to fall where it can be collected and removed. Yet another method of removing dust from baghouse surfaces is to force clean air in the direction opposite to normal flow.

Baghouses are designed using air-to-cloth ratios that are specific to the dust being removed. These ratios are given in terms of the flow rate of the gaseous stream being treated to the area of the cloth needed to treat that stream. These

Table 3.7 Air-to-cloth ratio for baghouses (US EPA [4])

Dust	Shaker/woven or reverse air/woven, $m^3/min\text{-}m^2$	Pulse jet/felt, $m^3/min\text{-}m^2$
Alumina	0.8	2.4
Asbestos	0.9	3.0
Bauxite	0.8	2.4
Carbon black	0.5	1.5
Coal	0.8	2.4
Cocoa	0.8	3.7
Clay	0.8	2.7
Cement	0.6	2.4
Cosmetics	0.5	3.0
Enamel frit	0.8	2.7
Feeds, grain	1.1	4.3
Feldspar	0.7	2.7
Fertilizer	0.9	2.4
Flour	0.9	3.7
Fly ash	0.8	1.5
Graphite	0.6	1.5
gypsum	0.6	3.0
Iron ore	0.9	3.4
Iron oxide	0.8	2.1
Iron sulfate	0.6	1.8
Lead oxide	0.6	1.8
Leather dust	1.1	3.7
Lime	0.8	3.0
Limestone	0.8	2.4
Mica	0.8	2.7
Paint pigments	0.8	2.1
Paper	1.1	3.0
Plastics	0.8	2.1
Quartz	0.9	2.7
Rock dust	0.9	2.7
Sand	0.8	3.0
Sawdust (wood)	1.1	3.7
Silica	0.8	2.1
Slate	1.1	3.7
Soap detergents	0.6	1.5
Spices	0.8	3.0
Starch	0.9	2.4
Sugar	0.6	2.1
Talc	0.8	3.0
Tobacco	1.1	4.0
Zinc oxide	0.6	1.5

ratios are multiplied by the flow rate of the gas stream to find the area of fabric needed for the baghouse. Table 3.7 gives the ratio of air to cloth for several types of dust and three types of baghouse.

Electrostatic precipitators are also used to remove particles from gaseous streams. In electrostatic precipitation, the gas stream passes through an electric field, which causes the particles to develop a negative charge. The charged particles then collect on electrically grounded plates. The effectiveness of electrostatic precipitators depends on the electrical resistance of the particles to be removed, on particle size, on the collection area of the precipitator, and on the volumetric flow rate of the gas being treated. The Deutsch–Anderson equation for estimating the fraction of particles that are collected in an electrostatic precipitator is

$$\eta = 1 - \exp\left(-\frac{WA}{Q}\right)$$

In this equation, W is the terminal drift velocity of the particles, A is the total collection area, and Q is the volumetric flow rate of the gas stream.

Condensation can sometimes be used to remove gaseous pollutants from a stream. The condensate can frequently be added to the process feed or be sold as a product.

Contact with aqueous streams is used to remove pollutants and to cool hot vapor streams. A wet scrubber is a common means of controlling gaseous and particulate air pollution. The efficiency of these systems depends on the degree of contact between the acid and gas streams, with extensive contact being desirable. The efficiency of acid gas removal depends strongly on the pH of the scrubbing liquid; the scrubbing liquid should be maintained at a pH in the 9–11 range.

Henry's law, which predicts equilibrium concentrations in dilute liquid and vapor phases, is applied when using wet scrubbing to remove a gas by dissolving it in the aqueous stream. Henry's law can be stated as follows:

$$y^* = Hx$$

In this equation, y^* is the mole fraction of pollutant in the gas phase that is in equilibrium with the liquid, H is the unitless Henry's law constant, and x is the mole fraction of pollutant in the liquid phase. The Henry's law constant is dependent on temperature; the solubility of a gas decreases with increasing liquid temperature. Compounds with high Henry's law constants are not effectively removed by contact with liquid. At equilibrium, no net mass transfer occurs, so pollution control equipment must be operated away from equilibrium conditions. Henry's law does not apply in high-concentration solutions or in situations where the pollutant reacts or dissociates in the liquid.

The Henry's law constant can be expressed in many different units. Sometimes the Henry's law constant is given in terms of molar concentration in the liquid phase instead of mol fraction in the liquid phase. Another form of Henry's law is

$$p_i = hx_i$$

In this equation, p_i is the partial pressure of compound i, h is the Henry's law constant, and x_i is the mol fraction of the compound in the liquid.

The Antoine equation can be used to evaluate vapor pressure at different temperatures. It is

$$\log_{10}(P) = A - \frac{B}{C + T}$$

In this equation, A, B, and C are constants specific to a temperature range and compounds, P is the vapor pressure of the compound, and T is the temperature in °C.

Venturi scrubbers are a type of wet scrubber that are used to remove particulate matter. In a Venturi scrubber, the stream to be treated passes through a Venturi tube. The gas stream must speed up as it passes the constriction in the middle of the tube. Water is sprayed into the stream either before or at the constriction in the tube. The particles and water combine, and the reduced velocity at the expanded section of the throat allows the now-contaminated water to drop out and be collected.

Wet scrubbers generate an aqueous waste stream that often must be treated before disposal or discharge.

Adsorption of gases onto solids is used in air pollution control. The solid can be activated carbon, silica gel, and alumina. Adsorbers are designed to have high surface area per unit volume. Activated carbon is useful for removing trace levels of organic contaminants in an air stream. Some carbon adsorbers are designed so that they can be regenerated using steam. Pollutants are sometimes recovered from the aqueous regeneration stream.

As with wet scrubbers, the use of adsorption to control air pollutants results in the generation of waste. Even adsorbers that can be regenerated become fouled or poisoned so that they are no longer usable, and the aqueous stream from the regeneration must be processed before it can be disposed of.

Incineration is another means of air pollution control. Complete combustion is necessary for destruction of organic compounds. It ranges in complexity from a simple flare to catalytic incineration. Auxiliary fuel is sometimes added to incinerators to ensure stable operation.

3.11 Odor Generation and Control

Sometimes a compound is released from a process or activity that is not a health or environmental concern but that is problematic because of its smell. Also, there are many compounds that are detectable by at least certain people at levels that are below health-based regulatory levels.

Total odor generation is proportional to the concentration of odor-causing compounds and the volumetric flow rate of emissions. Odor generation is measured in odor units. Human noses are the only instrument that can measure odor, as applied to nuisance odors and odor control. Of course, instruments that measure the concentration of compounds that create an odor nuisance exist, and these instruments can be used in cases where the compounds that cause a potential nuisance have been identified and can be measured. However, odor nuisances are usually caused by a number of compounds and can occur at concentrations so low they are difficult to measure.

The dilution threshold for an odiferous stream, expressed in terms of the number of volumes of fresh air needed to dilute a volume of odiferous air, is defined as the point at which one-half of the people can detect an odor in the diluted stream. It is abbreviated as ED_{50}. Odor concentrations are frequently expressed in terms of dilutions/threshold (D/T). A concentration of 1 D/T would represent a concentration detectable to half of the population. A sample at 5 D/T has a concentration equal to one-fifth of a sample at 1 D/T. Nuisance thresholds vary for different odor-causing compounds. The nuisance threshold for a typical composting facility is 5 D/T.

Many of the same techniques for reducing air pollutants apply to odor control. If practicable, the source of the odor should be eliminated. Sometimes the compound or compounds causing the odor can be captured and recycled to the process.

In some cases, an odor nuisance can be eliminated by dilution. Tall stacks are a means of dilution. In some dilution applications, less expensive and less unsightly mixed-flow fans that mix the odor-causing stream with fresh air and expel it upward at high speeds can be used instead of stacks. Masking, or adding a compound with a pleasant odor to a stream with an unpleasant odor, is another technique for reducing odor problems. Counteractants are compounds that, when added to a stream, offset the intensity or character of other odor-causing compounds so that upon perception the two odors cancel out each other. Counteractants are compound specific. Both counteractants and masking agents can only be used in situations where good contact with odor-causing compounds can occur.

At composting facilities, more compounds associated with odor problems are created during anaerobic conditions than during aerobic conditions. Low solid content, excessive moisture, and inadequate aeration all favor anaerobic conditions. Too much nitrogen results in ammonia releases, and too little nitrogen slows the rate of digestion.

Scrubbers can be used to control odor if the odor-causing stream or streams can be captured. Scrubbers alone do not generally prove effective for compounds that create an odor problem at very low concentrations.

Biofilters that adsorb odor-causing compounds and degrade them biologically generally have low operating costs, but they require a large area in which to operate.

3.12 Preventing Air Pollution

All the control strategies mentioned previously are added on to existing processes to reduce air pollution. Before any control system is applied, techniques for minimizing the creation of pollutants in the first place should be explored. These techniques can include process changes, changes in fuel, and implementation of good operating practices. An example of a process change would be the conversion of a process for making a chemical species to another process that has few air emissions. An example of a change in fuel would be to use coal with low sulfur content, rather than coal with high sulfur content, or to replace coal with natural gas, a less polluting fuel.

In order to minimize the creation of pollutants, the chemistry of the process responsible for their formation must be understood. An example is presented here for NO_x formation due to the combustion of fossil fuels. During combustion processes, organic fuel ideally would be converted to carbon dioxide and water, as illustrated in the following equation using methane as an example.

$$CH_4 + 2O_2 \rightarrow CO_2 + 2H_2O$$

However, impurities in the fuel, high temperatures, and imperfect conditions result in the release of unburned hydrocarbons, nitrogen oxides, sulfur dioxide, carbon monoxide, particulate matter, and metals. Nitrogen oxides are produced either by thermal fixation of nitrogen in the air with oxygen in the air (thermal NO_x) or by oxidation of nitrogen present in the fuel (fuel NO_x). The formation of NO via thermal processes can be described in two simplified steps, as follows.

$$N_2 + O \rightleftharpoons NO + N$$
$$N + O_2 \rightleftharpoons NO + O$$

The first reaction is the rate-determining step and requires high temperatures. The rate equation for this step, at temperatures greater than 2,800 °F, is

$$[NO] = k_1 \exp\left(-\frac{k_2}{T}\right)[N_2][O_2]^{1/2}t$$

In this equation, brackets indicate mole fraction, k_2 and k_1 are rate constants, T is the peak temperature in K, and t is the residence time of the reactants at the peak temperature. This equation shows that the formation of thermal NO_x increases exponentially with increasing temperature but is only weakly dependent on the concentration of oxygen and that the rate of formation of thermal NO_x can be reduced by reducing combustion temperatures.

The formation of fuel NO_x, on the other hand, is only weakly dependent on temperature. The formation of fuel NO_x from liquid fuels may be highly dependent on oxygen concentrations, so that limiting excess air has a dramatic effect on NO_x production. If fuel NO_x is responsible for the majority of NO_x emissions, a change to a lower-nitrogen fuel may be warranted.

Good operating practices include common sense measures, such as good housekeeping and proper maintenance. Regular inspection of equipment using simple leak detection devices, followed by prompt maintenance of equipment that is leaking at unacceptable rates, is an example of a good operating practice that reduces fugitive emissions.

Techniques for preventing the formation of pollutants have an advantage in that they result in greater material efficiency and/or less product loss.

References

1. David C. Cooper & Alley, F. C. (2011). Air pollution control: A design approach, (4th ed). Long Grove, IL: Waveland Press Inc.
2. Ranchoux, R.J.P. (1976). Determination of maximum ground level concentration. *Journal of the Air Pollution Control Association, 26*(11).
3. US EPA. (1970). Workbook of atmospheric dispersion estimates.
4. US EPA. (1990). OAQPS control cost manual. EPA 450/3-90-006.

Chapter 4
Solid Waste

4.1 Sources of Solid Waste

Sources of solid waste are as follows: residential; commercial establishments, industrial establishments; construction and demolition wastes; and agricultural operations and other municipal wastes (waste found on roads, waste from municipal waste treatment plant, etc.).

Residential waste includes food wastes, paper, cardboard, plastics, textiles, leather, yard wastes, wood, glass, metals, ashes, special wastes (e.g., bulky items, consumer electronics, used appliances), and household hazardous wastes. Food-based waste is often called garbage. Residential waste such as metals, glass, plastics, scrap rubber, leather, etc. is often termed rubbish.

Commercial waste comes from shopping centers, hotels, restaurants, markets, office buildings, etc.

Construction and demolition wastes consist of wood, concrete, metals, asphalt, roofing materials, dirt, etc.

Industrial wastes are generated during manufacturing, fabrication, construction, power generation, and at chemical plants.

Agricultural wastes can vary widely ranging from waste from farm manure residue to leaves and plants, feedlot wastes, and waste from slaughter houses.

Municipal wastes other than those described above can be non-point source wastes like dead animals or point source wastes or non-specific wastes from a wide variety of sources. Municipal waste treatment plants generate increasing amounts of solid waste especially solid sludge, which must be disposed of carefully.

4.2 Methods for the Disposal of Solid Wastes

The methods are as follows: Sanitary Landfill, Composting, Recycling, Incineration, and Energy Recovery. They are described below, with one section being devoted to each method.

A. Naimpally and K. S. Rosselot, *Environmental Engineering: Review for the Professional Engineering Examination*, DOI: 10.1007/978-0-387-49930-7_4, © Springer Science+Business Media New York 2013

4.3 Sanitary Landfill

A sanitary landfill can be defined as a method of disposing of refuse on land without creating nuisances or hazards to public health or safety, by utilizing the principles of engineering to confine the refuse to the smallest practical area, to reduce it to the smallest practical volume, and to cover it with a layer of earth at the conclusion of each day's operation, or at such more frequent intervals as may be necessary.

Although an acceptable method of solid waste disposal, it has not received wide public acceptance. There seems to be a general misconception of a sanitary landfill which results from the fact that the vast majority of solid waste disposal sites were open dumps. A recent survey has revealed that of an estimated 17,000 sites, only 6 % met the minimum requirements of a sanitary landfill. Between 1966 and 1970, many states passed legislation approving sanitary landfills and prohibiting open dumps. These actions are becoming more effective as a common state requirement is that a detailed engineering plan be developed and submitted to the appropriate state agency before a new sanitary landfill can be approved.

A regional system for solid waste management is generally preferred where one or more sanitary landfills serve several communities. The major reason is economic unit costs are lower because of the large volume of waste handled, and funds can be more easily provided.

Data on the quantity and characteristics of the solid waste produced in the communities served by the sanitary landfill must be available for both planning and design. Seasonal variation in solid waste produced should be clearly identified.

Getting public acceptance is one of the greatest problems in establishing a new sanitary landfill, primarily because most people equate this disposal method with a dump. A public education program is often required. A completed site can often be converted into a community asset—a golf course or playground.

The site chosen should provide enough capacity for at least a 20-year period. Ideally, the disposal site should be located in the center of the waste generation area, but this is usually impractical because it would be in a residential area.

Therefore, sites are generally near the fringe of the waste generation area. A site should have topographic and soil characteristics that will keep the amount of surface water infiltrating the landfill at minimum. Many of the problems with regard to surface water pollution can be avoided in this manner.

4.3.1 Site Selection

The typical site selection methodology involves considering the objective technical factors as well as the social and economic factors.

4.3.1.1 Objective Technical Factors

- The primary landfill requirements would involve the type of waste being disposed, the annual mass disposed of from the per capita solid waste data, and the volume of waste being disposed of at the site.
- Several possible sites must be considered, and data for these sites must be obtained: land-use maps, topographic maps, soil maps, highway maps, and bridge-loading maps. Any restrictions such as airports, floodplains, and wetlands must be considered.
- Hydrogeological considerations must be taken into account including seismic impact studies, presence of any landslide areas, and a good distance of at least a mile from drinking water sources must be considered. Soil borings must be taken to test the nature of the soil. Existing well logs, groundwater observation wells, and groundwater contour maps must be done as well because of the possible effect on groundwater recharge.
- Proximity to wastewater treatment facilities for leachate would be an important consideration as well as the availability of waste to obtain energy from landfill gases and treat the leachate on-site.

4.3.1.2 Social and Economic Factors

- Social factors include consideration of economic justice issues.
- Public input is required prior to siting, and important public concerns could be traffic to/from the site, air pollution, water pollution to groundwater, and public health concerns of neighbors.
- Cost estimate of the entire project must be made for all alternative sites: cost of land, roads, landfill, closure costs for site, on-going post-closure monitoring, etc. The costs of several alternative sites may need to be compared while considering all other social and technical factors.

4.3.2 Operation of the Landfill

Solid waste should be compacted on the site and then spread out on the ground. A layer of 6 in. of compacted soil must be spread out over the waste at the end of each day. Good maintenance should be provided for, and vector control may be necessary from time to time.

Storage of wastes at a site can lead to three possible consequences:

- Biological decomposition due to bacteria and fungi.
- Absorption of liquids as from food wastes and garden wastes.
- Contamination of one type of waste with another type.

Compacting solid waste into the form of rectangular blocks is called baling.
Volume reduction can be expressed by two formulas:
Percent volume reduction = (initial volume − final volume)/initial volume
Compaction ratio = initial volume/final volume

Burning of waste must be prohibited at the site. Also, recycling operations and salvage of materials must be conducted at other locations to avoid chaotic conditions.

Anaerobic decomposition of the solid waste occurs until long after the site has been closed. The principal gaseous products of decomposition are methane, nitrogen, CO_2, H_2, and H_2S. Studies on the quantity, rate, and composition of gas production have been extremely limited. During the early years, the predominating gas is CO, while during the later years, the gas is composed almost equally of CO_2 and methane. Methane is explosive, and various methods are used for controlling its movement. Differential settlement of the surface can result from different rates of reaction at various points in the landfill. Many of the above problems can be eliminated by utilizing the method of "aerobic landfill".

There are 5 phases of refuse stabilization within the landfill. In the first phase, the lag phase, aerobic bacteria become established and moisture builds up. In the second phase, the oxygen in the landfills is used by aerobic microorganisms. As a result, the temperature increases in the landfill. In the third phase, conditions are anaerobic and acids are formed, reducing the pH of the leachate. In the fourth phase, methane fermentation begins and the acids formed in the third phase are consumed. Methane, carbon dioxide, and water are formed along with sulfides. In the final, fifth phase, biological activity slows and very little gas is produced.

4.3.3 Water Balance in Landfills

The water falling as rainfall either enters the unit cell or becomes runoff. The net water entering the unit cell due to rainfall [1, 2]: (Fig 4.1).

$$W_R = (P - R) \text{ where } P = \text{Rainfall and } R = \text{Runoff}$$

In addition to water entering through rainfall, water entrained in the wastes is landfilled and water is used for dust control. In turn, water can leave through evaporation as water vapor, through consumption of water in the formation of landfill gas, and leave as leachate.

The difference between the water entering and leaving constitutes the net increase in water held within the landfill.

Thus:

$$\Delta S = W_S + W_C + W_R + W_T - W_G - W_L - W_E - W_{LE}$$

Fig. 4.1 Water balance in landfills

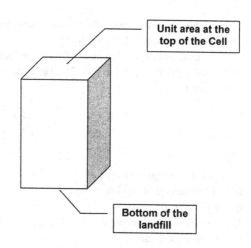

Unit area at the top of the Cell

Bottom of the landfill

where

ΔS	Increase in water held within a unit
W_S	Water entering through solid waste
W_C	Water entering through soil covering the cell
W_R	Water entering due to rainfall
W_T	Water entering through incoming treatment plant sludge
W_G	Water consumed in the formation of landfill gas
W_L	Water leaving as water vapor with the landfill gas
W_E	Water lost due to evaporation
W_{LE}	Water leaving from the bottom as leachate

The water balance for a unit area of soil used as cover above compacted solid waste in a landfill is

$$\Delta S_{LC} = P - R - ET - PER_{SW}$$

where

ΔS_{LC}	Change in amount of water held in storage in landfill cover
P	Amount of precipitation per unit area, in
R	Amount of runoff per unit area, in
ET	Amount of evaporation and transpiration, in
PER_{SW}	Water percolating into solid waste from the soil covering the landfill.

4.3.4 Leachate

The movement of leachate in a landfill can cause heavy metals and organics to move down. The cation exchange capacity (CEC) of a solid is equal to the number of milliequivalents (meq) of cations absorbed by 100 gm of soil. Clay liners prevent the passage of heavy metals.

The movement of leachate from the landfill into the ground water is an important environmental issue. The breakthrough time for leachate to penetrate in clay liner is [1, 2]

$$t = \frac{d^2\eta}{K(d+h)}.$$

t Breakthrough time (year)
d Thickness of clay liner (ft)
η Effective porosity
K Coefficient of permeability (hydraulic conductivity) (ft/year)
h Hydraulic head (ft)

Typical effective porosity values for clays with a coefficient of permeability in the range of 10^{-1}–10^{-8} cm/s vary from 0.1 to 0.3.

Organics found in the leachate can move with the gross leachate flow, subject to retardation. The retardation causes the organics to be held in the soil.

Where C_O and C_W are at equilibrium with each other, the *octanol-water partition coefficient* is defined as (1,2):

$$K_{OW} = \frac{C_O}{C_W}$$

C_O Concentration of chemical in the octanol phase, $\frac{mg}{L}$ or $\frac{\mu g}{L}$

C_W Concentration of chemical in the aqueous phase, $\frac{mg}{L}$ or $\frac{\mu g}{L}$

Soil-water partition coefficient, where X and C are in equilibrium with each other:

$$K_{SW} = \frac{X}{C}$$

X Concentration of chemical in soil, ppb or $\frac{\mu g}{L}$ and

C Concentration of chemical in water, ppb or $\frac{\mu g}{L}$

Organic Carbon Partition Coefficient, K_{OC} (1,2):

$$K_{OC} = \frac{C_{Soil}}{C_{Water}}$$

where
C_{Soil} Concentration of chemical in organic carbon component of soil (μg absorbed/Kg organic or ppb)

C_{Water} Concentration of chemical in water (ppb or $\frac{\mu g}{Kg}$)

K_{SW} $= F_{\text{OC}}K_{\text{OC}}$, where

F_{OC} Fraction of organic carbon in the soil (dimensionless)

Retardation Factor:
Retardation factor R is in general given by:

$$R = 1 + \frac{\rho}{\gamma}K_D$$

ρ Bulk density of soil in unsaturated zone
γ Porosity
K_D Distribution coefficient between the liquid phase and the solid phase in the soil.

Liners are used to protect the soil and groundwater around a landfill from contamination. The layers of liners are as follows:

- Several feet of protective soil layer
- A geotextile layer
- Several feet of sand or gravel leachate allocation layer
- Several mil. of geomembrane layers
- Several feet of compacted clay

The geomembrane and clay layers are relatively impermeable ones. The leachate generated is collected by a piping system placed during the construction of the landfill. The piping systems can consist of either percolated pipes or piped bottom for the landfill.

The leachate removed from the landfill can be dealt with in any one or more of the following manners:

- Sent to an integrated municipal wastewater treatment plant
- Treated on-site
- Disposed of, following on-site treatment or sent to the municipal waste water treatment plant following on-site treatment
- Leachate is evaporated
- Leachate is recycled through the landfill. This can increase the amount of methane generated within the landfill due to the increased acidity of the leachate.

After the landfill is completed, it begins to settle, with organic materials in the landfill being converted to leachate and landfill gas. About 30–35 % of the mass is lost in this manner, mainly in the first 5–8 years. Material placed above solids in the landfill causes it to be compacted as per the formula:

$$SW_P = SW_i + \frac{p}{a + bp}$$

where
SW_i Initial specific weight, lb_m/yd^3
SW_P Specific weight under pressure p, lb_m/yd^3
p Overburden pressure, lb_f/in^2
a, b Constants

After landfills are closed, continued monitoring of the groundwater, landfill air quality, and the soil vapor layer is needed.

4.3.5 Landfill Gas

Carbon dioxide produced in the landfill is heavier than air and go to the bottom of the landfill. Methane is poisonous and can also be explosive in mixtures with air. Methane can move in the horizontal direction through soil and lodge in the basements of homes and buildings close to the landfill. Therefore, barriers to the flow of methane must be constructed at the bottom and sides during the landfill construction phase. The landfill construction phase must also include a venting system for collecting the gas and managing it.

The organic matter in the solid waste undergoes anaerobic degradation through reactions of the type:

(Organic matter containing carbon, oxygen, nitrogen, hydrogen) + H_2O
$\rightarrow mCH_4 + nCO_2 + pNH_3 +$ degraded organic matter.

Unless collection mechanisms are in place, the landfill gas vents through the landfill cover to the atmosphere by diffusion and convective mechanisms.

4.3.5.1 Upward Movement of Landfill Gas to the Landfill Cover

The formula is [1, 2]:

$$N_A = \frac{D\eta^{4/3}(C_{A\,fill} - C_{A\,atm})}{L}$$

where

N_A Gas flux of compound A, $\frac{g}{cm^3-s}\left(\frac{Lb\,mol}{ft^2 d}\right)$

$C_{A\,fill}$ Concentration of compound A at the bottom of the landfill cover, $\frac{g}{cm^3}\left(\frac{Lb\,mol}{ft^3}\right)$

$C_{A\,atm}$ Concentration of compound A at the surface of the landfill cover, $\frac{g}{cm^3}\left(\frac{Lb\,mol}{ft^3}\right)$

L Depth of landfill cover, cm (ft)

D Diffusion coefficient, $\frac{cm^2}{s}\left(\frac{ft^2}{d}\right)$

Typical values for D for methane and carbon dioxide are $0.20\ \frac{cm^2}{s}\left(18.6\left(\frac{ft^2}{d}\right)\right)$ and $0.13\ \frac{cm^2}{s}\left(12.1\left(\frac{ft^2}{d}\right)\right)$

η Total porosity, $\frac{cm^3}{cm^3}\left(\frac{ft^3}{ft^3}\right)$

4.3.5.2 CO$_2$ Formation and Leachate

When CO_2 degasses from the leachate, it leads to an increase in Ca^{2+} ions and hardness, which can lead to equipment maintenance problems. The pH of the leachate also changes.

4.3.5.3 Use of Landfill Gas

The CO_2 and CH_4 in the landfill gas can be separated using membrane technology. The CH_4 can be used for generating electricity by burning the CH_4 and using the heat to generate steam which in turn can run the turbines. Otherwise, the CH_4-laden gas can be flared, if there is no use for it.

Good control systems need to be established for the control of landfill gas by means of the following: interceptor trenches, vents/flows, slurry wells, impermeable membranes at the bottom and the sides, sorptive compost, etc.

4.4 Recycling

Recycling and recovery are important from three perspectives:

- Generates incomes for the community or the recycler
- Saves natural resources
- Reduces the pressure on landfill

Several types of materials have been successfully recycled: paper, aluminum, copper, ferrous metals, plastics, tires, construction and demolition wastes, yard wastes, batteries, and oil.

4.4.1 Paper

Paper collected from recycling contains many other materials like rubber bands, clips, ink, coatings, etc. These have to be removed, and the paper should be converted to the required end use. Waste paper converted to newsprint or other type of paper needs to be chemically pulped after being deinked by one of the several available deinking processes. The post-pulping process is similar to that of regular papers. Fiber lengths are cut short due to the chemical and mechanical treatments of the waste paper and it is not possible to make high quality paper from 100 % recycled paper.

Paper can also be recycled without the deinking process by being converted into paper boards, lining boards for home construction, etc. Finally, paper can make an excellent fuel for incineration due to its high calorific value.

4.4.2 Aluminum and Copper

Aluminum and copper are the principal non-ferrous metals which can be recycled. Aluminum can be separated from other metals either by a gravity separator or by a magnetic one. The high cost of electricity to manufacture virgin aluminum from bauxite has made aluminum an excellent candidate for recycling. High-purity aluminum recycle streams can be used as a good substitute for the virgin aluminum, while the lower-grade streams are useful for the manufacture of alloys. One of the sources of recycled copper is electrical wires from which insulation is removed and the copper is obtained by melting the wire.

4.4.3 Ferrous Metals

Iron and steel are recycled from steel cans (used for food) and from automobile and other heavy scrap. The steel cans contain tin which must be removed if the end product is steel; otherwise, some tin can be tolerated for end uses other than steel making. Automobile and heavy scrap are recycled using technology that is presently mature in the USA. Steel is obtained by melting the scrap steel and subsequent purification steps.

4.4.4 Glass

Glass is a material that occupies space in landfills and consumes energy for production. The raw material of glass (silica) is cheap and abundant. Glass can be separated using the following methods: froth flotation, dense media separation, and color sorting (ref: [5]). Clear melted glass ("cullet") can be mixed during the preparation of virgin glass to reduce furnace temperature and energy. Colored glasses need to be sorted for reuse. Other end uses of recycled bulk melted glass are glassphalt (similar to asphalt) for roadwork, fiberglass, etc.

4.4.5 Plastics

Plastics, which are essentially polymers made from organic materials, are non-biodegradable, have a low specific weight, and thus occupy large volumes in landfills. (The large variety of plastics and the increasing number of uses have made it imperative that better sorting methods, less number of plastics, and perhaps more limited use of them be implemented in the future to mitigate the rise in landfill use). There are several varieties of plastics: polystyrene (SPI code 6-PS), polyethylene terephthalate (1-PETE), high-density polyethylene (2-HDPE), etc.

Polystyrene is used in items such as foam cups and plates, polyethylene terephthalate is used in various objects such as soft drink bottles, and high-density polyethylene is used for many things, including milk bottles. Each one should be sorted out and has its own difficulties if the objective is to use it as a substitute for the original virgin plastic. The major impediment to greater recycling are the large variety of plastics, the transportation costs, and the large volume to mass ratio that makes recycling uneconomical and impractical at times.

4.4.6 Used Tires

Used tires are an excellent fuel for waste-to-energy plants besides being used for minor uses like roadway barriers.

4.4.7 Construction and Demolition Wastes

Wood, asphalt, concrete, and metal can be recovered and recycled in some manner. Wood can be recycled only if in good condition; otherwise, it can be incinerated after processing or used as landscape material. Asphalt and concrete wastes can be reused as a road base. The metals from construction and demolition materials, if sorted well, are the only category that can be relatively easily recycled.

4.4.8 Yard Waste

Yard waste can be processed as compost (see Sect. 4.5) or used as landfill cover.

4.4.9 Batteries

Automotive and equipment batteries contain lead, a large fraction of which is recycled due to the toxicity of lead. Home-use batteries contain toxic chemicals like mercury and cadmium, which can cause leachate from landfills to pollute groundwater with these heavy metals. It is desirable that more home-use batteries be recycled and/or alternative methods to manufacturing domestic batteries be found.

4.4.10 Waste Oil

Waste oil can be collected at many places and refined for recycle.

4.5 Composting

Composting is the process of converting organic solid waste into humus or particle-like material by means of aerobic biological degradation. Generally, the organic solid waste, which must be separated from MSW, is mixed with sewage sludge and then degraded by means of one of several processes. Prior to the composting process, it may be necessary to sort and separate non-degradable solids like glass, aluminum, etc. This is followed by shredding and pulverizing the material so that particles are small in size for effective biological action.

Three common processes are:

Windrows: A conical pile which is about 5 ft high and 10 ft long. The pile can be periodically turned by mechanical means. This method can be done either in an open field which needs more land, or in enclosed facilities which are artificially aerated.

- An aerated pile is another aerobic method
- A flow of organic solid waste through an enclosed vessel with air injection

The processed composed is generally dried, screened, and either granulated or palletized for marketability.

For the composting process, it is important to keep the C:N ratio at around 22:1. The temperature is kept between 50 and 55 °C for the first few days and between

55 and 60 °C thereafter. Moisture control is also important with the desirable moisture being around 55 %.

Microorganisms of various kinds are required for the process: bacteria, fungi, actinomycetes, etc. Initially, the mesophilic bacteria converts the organic matter to CO_2 and increases the temperature to 40 °C. Thereafter, the thermophilic bacteria act on it and increase the temperature to 60 °C. If sufficient microorganisms are not present in the solid waste, it may be necessary to add them. The final product of the composting process can be used as soil amendment.

4.6 Incineration and Energy Recovery

Incineration of solid wastes, along with energy recovery during the process, has gained popularity in recent years in the USA. In several other countries including Japan, it is the dominant mode of solid waste disposal. Solid waste can be incinerated "as is" or can be burned after removing all non-combustible items through shredding and separation processes. The latter fuel is called RDF or refuse-derived fuel. Solid waste has a calorific value that is about a third of that of coal.

The Dulong formula can be used to obtain the value of solid waste. When the chemical formula of the solid waste is known, the energy content in $\frac{BTU}{lb_M}$ is given by [2, 3]:

$$E, \frac{BTU}{lb_M} = 124\,C + 610\,(H_2 - 1/8\,O_2) + 40\,S + 10\,N$$

where C, H_2, O_2, S, and N denote the weight percent of carbon, hydrogen, oxygen, sulfur, and nitrogen.

The combustion of solid waste can be done in different types of incinerators: fluid bed, rotary, multiple hearth, etc.

Water can be heated or converted into steam using the heat generated during the process. Steam, if generated, could be used to run turbines and generate electricity. Alternatively, the steam could be used for heating residences in cold climates and the waste water from the heating process, in turn, can be used as process water in industry.

A significant residue of ash will be left as a result of the combustion process. The fly ash, which leaves with the gases leaving the furnace, may contain toxic substances. The bottom ash, left behind in the furnace, may contain heavy metals. The ash from the combustion process used to be disposed of in separate, secure landfills. Alternatively, it can form a component of specific cements. The combustion process needs a good air pollution control system to contain >99 % of the ash, SO_2, chlorites, metal, etc. during the post combustion process.

4.7 Solid Waste Collection and Storage

Solid waste collection and storage accounts for more than 60 % of the costs of managing solid wastes. This is due to the distributed nature of solid wastes in geographical terms, the distances over which the wastes are hauled and the manpower required for the purpose.

The materials collected from residences and commercial establishments can be either hauled directly to the landfill or incineration facility, or can be taken to an intermediate point from which it can be hauled by larger trucks. The intermediate point in the latter case can be either a non-storage one where the solid waste is directly transferred to the larger trucks or a storage facility where one or two days of solid waste is temporarily stored. In the non-storage intermediate facility, there are two levels, with the recipient trucks being on the lower level for convenience of transfer through a chute.

Due to the high cost of collection and storage systems, many mathematical models have been developed for solid waste collection systems. The collection system can be described in 5 phases:

Phase 1: From the home to the curbside can—this is the simplest part of the process.

Phase 2: Collection of solid waste from the roadside can to the truck. It can be done either by mechanical means by the truck, or by a crew varying from 1 to 4 persons.

Phase 3: Movement of the collection vehicle through the streets as it collects solid waste and reaches either the intermediate point or the final destination of the solid waste. Several mathematical models and software are available for the optimum (in cost terms) route for this purpose.

Phase 4 (if applicable): Intermediate point for the solid waste (either of the storage type or the non-storage variety).

Phase 5: Long haul from the intermediate point to the final destination.

For phase 3, several rules-of-thumb are available:

1. Starting collection point must be close to the originating garage.
2. Avoid rush hours.
3. For sloped roads, collect when moving downhill.
4. Clockwise turns must be done on blocks.
5. Do not overlap paths. Long, straight paths should be followed.

Optimization techniques use systems of nodes and links to create the optimum path.

For phases 4 and 5, optimization involves utilizing the closest destination to the maximum extent, i.e., haul as much as it can be maximally received by this site. From here on, move on to the next further site by conveying the maximum receivable at that site. Continue until all of the solid waste is hauled away.

Trucks used for solid waste collection, whether front-loading, side-loading, or back-loading, are generally equipped with compactors.

4.8 Regulations

4.8.1 National Environmental Policy Act

The National Environmental Policy Act (NEPA), 1969 directs all federal agencies and projects governed by the federal agencies to have an Environmental Impact Statement and use a matter of proper procedure. The White House's Council on Environmental Quality (CEQ) was also set up by the NEPA [4].

This Act is thought to be the most important environmental legislation in the United States during the post World War II period. The Act along with regulations and formal executive orders makes sure that all public decision making occurs in a fair manner with respect to the environment. The Act has also influenced environmental legislation in many other countries.

The national goals from section 101 of the Act is to make each generation fulfill responsibilities as a trustee for the environment for subsequent generations; assure that all Americans are safe, healthy, and productive, and esthetically and culturally have pleasant surroundings; attain the widest range of beneficial uses of the environment; preserve important historical, cultural, and natural aspects of the national heritage, enhance the quality of renewable measures; and obtain maximum attainable recycling of depletable resources.

Section 102 of the Act states all agencies of the US government be required to utilize an interdisciplinary environmental impact assessment (EIA) method. Federal agencies are required to identify environmental amenities and give them appropriate consideration when planning actions that affect the environment. An environmental impact statement (EIS) must be prepared, in some cases. The CEQ coordinates federal environmental efforts and works closely with agencies and other White House offices in the development of environmental policies and initiatives. Along with the CEQ, the Environmental Protection Agency (EPA) was established in 1970. The EPA consolidated the diverse environmentally related units scattered across the federal government.

4.8.2 Comprehensive Environmental Response, Compensation, and Liability Act

Enacted in the aftermath of the Love Canal incident, the Comprehensive Environmental Response, Compensation, and Liability Act (CERCLA) has several objectives: respond to the presence of inactive hazardous waste sites; provide source of funds (Superfund) for cleaning up these sites; provide a means of determining liability; and establish rules for these sites. The Superfund Amendments and Reauthorization Act (SARA) amended CERCLA and added emphasis on the cleanup of hazardous waste sites and how the costs are to be handled.

The national priorities list (NPL) is an important part of CERCLA. Sites are placed on the NPL based on groundwater migration, surface water migration, soil exposure, and air migration of actual or potential releases of hazardous substances. The NPL ranking system was revised in 1990 in response to the SARA amendments of 1986. The national contingency plan (NCP) determines the remediating actions, and the priorities for NCP are based on the need to protect human health and the need to achieve the highest level of cleanup.

4.8.3 Resource Conservation and Recovery Act

The goal of the Resource Conservation and Recovery Act (RCRA), enacted in 1976 and amended in 1984, is to set a national policy of reducing or eliminating the generation of hazardous waste. The 1984 amendments establish a ban on disposal of nonhazardous liquids in hazardous waste landfills, minimum technological requirements for hazardous waste surface impoundments and landfills, and corrective action for any release from hazardous waste treatment, storage, and disposal (TSD) facilities. EPA has set out detailed regulations for the design of different types of TSD facilities. An important provision of RCRA is the ban on land disposal of untreated hazardous wastes. Private citizen lawsuits are allowed as an important tool by RCRA. The RCRA amendments also brought in small generators of hazardous waste.

RCRA came a few years after the Resource Recovery Act (RRA) of 1970 encouraged recycling of solid wastes and the recovery of energy as the primary means of solid waste management.

References

1. Tchobanoglous, G. et al. (1993). *Integrated solid waste management*. NJ: McGraw-Hill.
2. NCEES, *FE Reference Handbook*. (7th edn). (2005).
3. Vesilend, P. A., Rimer, A. E. (1981). *Unit operations in resource recovery engineering*. NJ: Prentice-Hall, Inc.
4. Sullivan, T. (2005). *Environmental law handbook*. (18th edn), Washington, D.C: Government Institutes.
5. Conway, R. A., & Ross, R. D. (1980). *Handbook of industrial waste disposal*. NY: Van Nostranol Rheinhold.

Chapter 5
Hazardous Waste

5.1 Definition and Characterization of Different Types of Hazardous Waste

The Resource Conservation and Recovery Act (RCRA) and, to a smaller extent, the Toxic Substances Control Act (TSCA) are the two main sources of regulation of hazardous wastes in the United States. Generators, transporters, and treatment, storage, and disposal facilities (TSDFs) are regulated under RCRA in a "cradle-to-grave" management system. States have the right to develop their own RCRA programs, but state-based RCRA programs must be consistent with and at least as stringent as the federal program.

Land disposal restrictions (LDRs) establish concentration- or technology-based treatment standards for most of the hazardous wastes that are regulated under RCRA and that are destined for land disposal. Wastes for which an LDR has been established or which are expressly restricted from land disposal are known as restricted wastes. Wastes that are eligible for a case-by-case extension, a no-migration exemption, or a national capacity variance can be land disposed without meeting treatment standards, provided that they are disposed in a landfill that meets minimum requirements. Waste analysis, notification, and record-keeping requirements apply to all restricted wastes, regardless of whether they have received an exception to LDR requirements. These requirements apply to hazardous waste generators, transporters, and TSDFs as the waste makes its way from generation to disposal.

A waste analysis plan (WAP) documents the procedures used to obtain a representative sample of the waste and to conduct a detailed chemical and physical analysis of the representative sample. A WAP is required for all TSDFs, as well as generators who treat their waste on-site in tanks, containers, and containment buildings. Waste analysis information gathered by generators must be forwarded to subsequent TSDFs.

RCRA regulations are labyrinthine, and this section will attempt a brief summary only. To begin, RCRA defines hazardous waste as either "characteristic" or "listed."

A. Naimpally and K. S. Rosselot, *Environmental Engineering: Review for the Professional Engineering Examination*, DOI: 10.1007/978-0-387-49930-7_5,
© Springer Science+Business Media New York 2013

Characteristic hazardous wastes, or D wastes, are wastes that are not otherwise listed and that exceed threshold properties for at least one of four characteristics: (1) ignitability, (2) corrosivity, (3) reactivity, and (4) leachability of specific listed compounds at concentrations greater than threshold levels (the threshold levels are intended to reflect toxicity).

A waste is considered ignitable if it

- has a flash point less than 140 °F, unless it is an aqueous solution containing less than 20 % alcohol by volume
- is a non-liquid that burns vigorously and persistently and that can ignite through friction or through absorption of moisture, or can ignite due to spontaneous chemical reactions
- is an ignitable compressed gas
- is an oxidizer

A waste is considered corrosive if it

- is aqueous with a pH less than or equal to 2.0 or a pH greater than or equal to 12.5
- is a liquid that corrodes carbon steel (grade SAE 1030) at a rate greater than 0.250 in/yr

A waste is considered reactive if it

- is normally unstable and undergoes violent physical or chemical change without detonating
- reacts violently with water
- forms a potentially explosive mixture when wetted with water
- can generate harmful gases, vapors, or fumes when mixed with water
- is a cyanide- or sulfide-bearing waste that can generate harmful gases, vapors, or fumes when exposed to pH conditions between 2.0 and 12.5
- will detonate or generate an explosive reaction when subjected to a strong initiating source or when heated in confinement
- is readily capable of detonation at standard temperature and pressure
- is an explosive listed as Class A, Class B, or "forbidden"

Leachability is an important characteristic of waste because it determines the likelihood of migration of toxic compounds after disposal. A waste that is characterized as hazardous because of its leachability is one whose acid extract contains one of about 40 listed compounds at a concentration greater than that of compound's threshold concentration. The procedure for determining leachability is called the toxicity characteristic leaching procedure or TCLP. The TCLP is intended to mimic sanitary landfill conditions that would result in contaminated landfill leachate. This category of characteristic wastes is generally referred to as "toxicity characteristic" hazardous wastes.

Wastes that are designated hazardous because of their characteristics (as opposed to listed hazardous wastes) become non-hazardous only if they are treated so that their hazardous characteristics are destroyed.

Listed wastes include wastes that contain certain toxic compounds ("acute hazardous wastes" are on the P list, "toxic wastes" are on the U list), wastes that are generated from specific sources (the K list), and wastes that are generated by non-specific sources (the F list).

A number of specific wastes have been excluded from the RCRA definition of hazardous, whether or not they meet any of the requirements of hazardous waste. These include domestic sewage, nuclear waste, household waste, mining over-burden, and wastes associated with the exploration, development, or production of crude oil and natural gas.

RCRA regulations govern the storage, handling, and disposal of hazardous wastes. These regulations include many reporting and record-keeping require-ments in addition to mandating storage, treatment, and handling standards.

Radioactive wastes are regulated under the Atomic Energy Act (AEA). Mixed wastes are defined as waste mixtures that contain both radioactive material subject to the AEA and hazardous waste subject to RCRA regulations.

5.2 Sampling and Measurement Methods for Hazardous Waste

Waste analysis is necessary in order to determine whether a waste is hazardous and to manage it properly. There are often several methods or combinations of methods that can be used to meet waste analysis requirements. Preferably, waste analysis requirements are met by conducting sampling and laboratory analysis because sampling and analysis is more accurate and defensible than other options. Acceptable knowledge (process knowledge and knowledge from previous waste analyses when circumstances have not changed) is an alternative to sampling and laboratory analysis and can be used to fulfill all or part of the waste analysis requirements.

Sampling of waste streams is complicated by the fact that these streams are not generally constant throughout a process or across time. Multiple samples are taken to account for this. The location and timing of sampling can be either authoritative (based on process knowledge) or random. A grab sample is one that is taken at a particular point and location, while a composite sample is a sample that combines many separate samples for analysis.

For liquid waste streams, samples are collected using composite liquid waste samplers (coliwasas), weighted bottles, and dippers. A coliwasa is most appro-priate when sampling free-flowing liquids and slurries in drums, shallow tanks, pits, and similar waste containers. The coliwasa consists of a glass or metal tube equipped with an end closure that can be opened and closed while the tube is submerged in the material to be sampled. If samples are to be taken from areas with limited accessibility, a weighted bottle or dipper may be used instead.

Triers, thieves, and augers are used to collect samples from solid waste streams and sludges.

A thief consists of two slotted concentric tubes and can be used to sample dry granular wastes whose particle diameter is less than one-third the width of the slots in the thief. A pointed tip in the thief's outer tube allows penetration of the waste that is being sampled. The inner tube can be rotated to open and close the sampler so that a sample can be collected from the waste stream before the thief is withdrawn.

A trier resembles a tube that has been cut in half lengthwise. The trier's sharpened tip allows penetration of the tube into adhesive solids and allows granulated materials to be loosened. Triers can be used to sample solid wastes whose particles have a diameter less than one-half that of the trier.

Augers consist of sharpened spiral blades attached to a hard metal central shaft and are used to sample hard or packed solid wastes. Dippers and shovels are also used to sample loose solid wastes.

If a tank is well-mixed, samples can be drawn off at side taps.

Samples must be preserved so that losses are prevented. Examples of sample preservation include the addition of sodium thiosulfate to inhibit organochlorine reactions, refrigeration, and selection of appropriate container materials and seals. In addition to preservation requirements, the EPA has established standardized holding times, based on the chemical constituent of interest, that must be met to assure the viability of analytical data from samples.

Quality control measures include preparation of field blanks, which consists of filling a sampling container in the field with a substance such as deionized water and any preservative that was used in the preparation of the actual sample. Airborne contaminants are an example of contamination that is revealed by field blanks. Trip blanks are sampling containers that are filled with deionized water and any preservative, then capped and taken unopened on the sampling trip. A contaminated trip blank can indicate contaminated sampling containers or contamination in the area where the trip blank was prepared. Equipment blanks are used to test for contaminated equipment and are taken by rinsing sampling equipment with deionized water and capturing the rinsewater. Split samples are two containers of the same sample and are used to determine whether analytical methods are producing consistent results. Field duplicates are samples that were taken in the same place at the same time and reveal consistency in waste composition and sampling methodologies.

For sampling of soils and solids, which is complicated by the fact that soil is an inhomogeneous medium, errors in sampling are likely to outweigh errors in the analysis. Data quality objectives have been set for sampling of soils and solids. These objectives are more stringent for planned removal and remedial response operations than they are for emergency clean-up operations, which are in turn more stringent than the data quality objectives for preliminary site investigation.

Once a sample is taken, it must be prepared for analysis. This can consist of extracting the analytes of interest, adjusting the physical properties, removing potential interferences, and/or concentrating analytes of interest. The determination of compounds of interest depends on the sample being analyzed. Organic compounds are often analyzed using gas chromatography (GC) or gas chromatography followed by mass spectroscopy (GC–MS). Metals are typically analyzed using atomic absorption spectroscopy (AAS) or inductively coupled argon plasma (ICAP) spectroscopy. Preparation and determination methods are prescribed by the EPA and are based on the compounds to be analyzed.

Opportunities for contamination exist during preparation and determination, and analytical laboratories must include method blanks to check for contamination of the sample at the laboratory. Other quality control procedures at laboratories include spiking samples with a known quantity of the compound of interest, analyzing a certified reference material, and analysis of duplicate samples to ensure that results are repeatable.

5.3 Storage, Collection, and Transportation Systems for Hazardous Waste

Hazardous waste is usually stored in containment buildings, tanks, and containers. Tanks are frequently used for waste treatment as well as for storage.

Design requirements for tanks depend on whether they are new or existing. The integrity of tanks must be regularly assessed, and they must be inspected for leaks. Tanks must be fitted with secondary containment and leak detection so that if failure occurs, hazardous waste is prevented from reaching the environment. Secondary containment has to take the form of an external liner, a vault, or a double wall. Other containment forms can be used if a variance is applied for and approved.

External liners must be large enough to contain 100 % of the volume of the tank's contents. Sometimes a single liner will surround more than one tank. If precipitation can gather within the liner, the volume must be large enough to hold the volume of the tank(s) plus the volume of a 25-year 24-h precipitation event. An example of a liner is a concrete-lined circular berm around a tank.

A vault is a lined underground chamber in which the tank rests or a closed aboveground building. Vaults must be large enough to contain the contents of the tank they enclose. A double-walled tank is a tank within a tank with a leak detection device that operates in the space between the tank walls.

If a tank wall is breached, the mass flow rate of liquid through the hole is given by

$$Q = CA_{\text{hole}} \sqrt{2\rho P_{\text{gauge}}}$$

In this equation, A_{hole} is the area of the hole, C is the discharge coefficient, the value of which is usually between 0.51 and 0.61, ρ is the density of the liquid, and P_{gauge} is the gauge pressure within the tank at the hole.

A container is any portable device in which a material is stored, transported, treated, disposed of, or otherwise handled. Fifty-gallon drums are an example of a hazardous waste container, but hazardous waste containers also include rail cars and test tubes. Containers are used for storage, transport, and disposal. Empty containers must in some cases be managed as hazardous waste. Special conditions must be met before containers are considered to be "RCRA-empty."

Containers at TSDFs must be placed in a storage area that will contain their contents in the event of a failure. This secondary containment usually consists of a concrete pad with a curb that drains to a sump.

Containment buildings are considered to be a hazardous waste management unit. A containment building is a completely enclosed structure (i.e., possessing four walls, a roof, and a floor) that houses an accumulation of non-containerized waste. The floor, walls, and roof must be constructed of man-made materials possessing sufficient structural strength to withstand movement of wastes, personnel, and heavy equipment within the unit. Fugitive emission controls are required. If the building contains wastes consisting of free liquids, passive secondary containment and leak detection equipment similar to that of storage tanks are required.

Transporters of RCRA hazardous waste are required to obtain an EPA identification number for the waste, comply with the manifest system, and deal with hazardous waste discharges. These regulations incorporate and require compliance with Department of Transportation provisions for labeling, marking, placarding, proper container use, and discharge reporting.

5.4 Minimization, Reduction, and Recycling of Hazardous Waste

Environmental management systems (EMSes) are adopted by facilities wishing to apply a systematic approach to environmental management. These systems are similar to quality management systems in that they require training, review, and documentation of procedures and processes.

An EMS does not require reductions in waste generation or reductions in the release of pollutants, nor does it require that less toxic materials be used. What an EMS does require is that each activity that has an impact on the environment be evaluated, and this evaluation must be documented. While EMSes do not require reductions in waste generation, such reductions are often the result of the implementation of an EMS.

Many people are familiar with ISO 9000, the International Standards Organization's quality management system standards. Companies can become ISO 9000 certified by applying for certification and passing an independent audit. The EMS parallel to the ISO 9000 standards is the ISO 14000 standards.

Facilities in the United States are not required to implement an EMS except in special circumstances. For example, federal facilities may be required to have an EMS, and sometimes an EMS is prescribed as part of a regulatory enforcement action.

The waste management hierarchy places waste management strategies in the following order, from highest preference to lowest:

1. Source reduction
2. In-process recycling
3. On-site recycling
4. Off-site recycling
5. Waste treatment to render the waste less hazardous
6. Secure disposal
7. Direct release to the environment

Source reduction is a change in operating practices, processes, or materials that results in a reduction in the amount of hazardous substance released to the environment or to a waste stream.

The costs associated with waste generation have risen faster than the rate of inflation. Many chemical plants and refineries were constructed decades ago when waste generation was less expensive. Sometimes, waste minimization or pollution prevention alternatives that were not economically attractive at an earlier date have become optimal due to the increasing expense of waste generation.

5.5 Treatment and Disposal Technologies for Hazardous Waste

Wastes that are designated as corrosive because they are at the extreme end of the pH scale can be neutralized. Basic solutions are neutralized by adding (in the form of a strong acid) the same number of hydronium ions as there are hydroxide ions in the solution. Acidic solutions are neutralized by adding (in the form of a strong base) the same number of hydroxide ions as there are hydronium ions in the solution. The amount of hydronium or hydroxide ions needed to neutralize a given amount of solution can be determined if the solution's pH is known, as follows:

$$pH = \log\left(\frac{1}{[H^+]}\right)$$

Solve for $[H^+]$ and multiply by the volume of the solution to get the moles of hydroxide ions required for neutralization

$$V[H^+] = \frac{V}{10^{pH}} = n_{OH, req}$$

The formula for calculating the number of moles of hydronium ions needed to neutralize a basic solution of known pH is similar.

Once the number of moles is found, the number of hydronium or hydroxide ions donated by the substance being used to neutralize the solution and that substance's molecular weight can be used to determine the mass of neutralizer required. For example, sodium hydroxide (NaOH) is commonly used to neutralize acidic wastes. Sodium hydroxide donates one hydroxide ion per molecule when it dissociates, so if dry sodium hydroxide is used to neutralize an acidic solution, the mass of sodium hydroxide can be found by multiplying the moles of hydroxide ions required by the molecular weight of sodium hydroxide.

Treatment standards are often given in terms of destruction and removal efficiency, which is calculated using the equation.

$$DRE = \frac{W_{in} - W_{out}}{W_{in}} \times 100\%$$

In this equation, W_{in} is the mass flow rate of a specific compound into the treatment device, and W_{out} is the mass flow rate of emissions of that compound from the treatment device.

Incinerators are sometimes used to destroy some of the constituents found in hazardous waste. The degree of destruction of waste in a rotary kiln incinerator depends in part on how much time the waste spends in the kiln. The kiln's residence time depends on the kiln's shape, the angle at which it is tilted, and the speed at which it is rotating, as follows:

$$t = \frac{19L}{DSN}$$

In this equation, t stands for time in minutes, L is the internal length of the kiln in meters, S is the kiln rake (the slope of the kiln) in cm/m, D is the diameter of the kiln in meters, and N is the rotational speed of the kiln in rotations per minute.

Poor combustion in an incinerator results in excessive emissions, and there are minimum requirements for the combustion efficiency of incinerators. The combustion efficiency of an incinerator is given by

$$CE = \frac{[CO_2]}{[CO_2] + [CO]} \times 100\%$$

In this equation, $[CO_2]$ and $[CO]$ represent the concentrations of carbon dioxide and carbon monoxide, respectively, in the incinerator's exhaust gas.

The feed to an incinerator must be a fuel of high enough quality to maintain efficient combustion. If the hazardous waste is not a high-quality fuel, it is incinerated with fuel that brings the combustion efficiency to acceptable levels. The amount of fuel required is determined by the heat deficiency of the waste stream and the fuel's heating value, as follows:

$$\text{quantity of fuel required} = \frac{\text{heat deficiency}}{\text{fuel heating value}}$$

Large heating values indicate high quality fuels.

In complete combustion, all of the carbon atoms present in a waste are converted to carbon dioxide and all of the hydrogen atoms present are converted to water. A compound's higher heating value, HHV, is that compound's heat of combustion when the products of combustion exit the reaction below the boiling point of water. A compound's lower heating value, or LHV, is the compound's heat of combustion when the products of combustion exit the reaction above the boiling point of water.

The heating values of a compound can be determined using the balanced combustion equation for that compound, along with the heats of formation for all the compounds except oxygen in the balanced equation. For the case of benzene, for example, the balanced combustion equation is

$$C_6H_6 + 7.5O_2 \rightarrow 6CO_2 + 3H_2O$$

The sum of the heats of formation for one mole of benzene, six moles of carbon dioxide, and three moles of liquid water divided by the molecular weight of benzene results in benzene's HHV. The sum of the heats of formation for one mole of benzene, six moles of carbon dioxide, and three moles of water vapor divided by the molecular weight of benzene results in benzene's LHV.

Most hazardous waste streams contain more than one hazardous chemical. When incineration is used as the waste treatment technique, the incinerator must be operated such that the most difficult-to-incinerate constituent is destroyed to acceptable levels. A compound's incinerability index can be used to determine which compound in a stream is going to be the most difficult to incinerate. It is calculated using the formula

$$I = C + \frac{100\,\frac{kcal}{g}}{H}$$

In this formula, C is the mass percent of the compound and H is its heating value.

Land disposal is placement of waste into a landfill, surface impoundment, waste pile, injection well, land treatment facility, salt dome formation, salt bed formation, underground mine, or underground cave. Hazardous waste land disposal units are required to perform groundwater monitoring, provide financial assurance, and meet closure and post-closure requirements.

New hazardous waste piles, landfills, and surface impoundments and expansions of existing waste piles, landfills, and surface impoundments are required to have a double liner, a leak detection system, and a leachate collection and removal system (LCRS). The double liner system consists of a top liner to prevent migration of hazardous constituents into the liner and a composite bottom liner consisting of a synthetic membrane and three feet of compacted soil material. The LCRS must collect liquids in a sump and subsequently pump out those liquids. In addition to the performance and design requirements, the LCRS must be located between the liners immediately above the bottom composite liner.

Surface impoundments are similar to landfills, but surface impoundments are intended for short-term storage or treatment while landfills are intended for final disposal. The difference between a surface impoundment and a tank is that the walls of a surface impoundment are supported by earthen materials, while a tank can maintain its structural integrity without earthen support if it is filled to capacity. In addition to a double liner, a leak detection system, and an LCRS, surface impoundments must establish a site-specific leachate flow rate, called the action leakage rate (ALR), to indicate when it is not functioning properly.

Waste piles are non-containerized piles of non-flowing hazardous waste. They are also used for temporary storage and treatment. Groundwater monitoring under waste piles is not required if the waste pile is (1) located inside or under a structure and does not receive free liquid, (2) protected from surface water run-on, (3) designed and operated to control dispersal of waste, and (4) managed to prevent the generation of leachate. The difference between waste piles and containment buildings is that containment buildings are not land disposal units. Containment buildings are designed with a containment system rather than a liner and leak detection system. A second LCRS above the top liner is required for waste piles.

Landfills are intended for final disposal. They require a double liner with an LCRS between liners and above the top liner, a leak detection system, and the development of an ALR. In addition, landfills must have storm water run-on and runoff controls to prevent migration of hazardous constituents for at least a 25-year storm, as well as a cover to prevent wind dispersal. Waste is often disposed of in hazardous waste landfills inside a container. To prevent significant voids that could cause the cover to collapse when the containers erode and to preserve available hazardous waste landfill capacity, containers placed in a landfill must be either at least 90 % full or have their volume reduced.

In land treatment units, hazardous waste is applied to the surface of soil or mixed with the top layer of soil so that it will be microbially and/or photolytically destroyed or immobilized. The effectiveness of land treatment must be demonstrated by field-testing or laboratory tests. Hazardous waste can only be placed in a land treatment unit if the waste will be rendered nonhazardous or less hazardous. Thus, land treatment units are limited to primarily organic wastes. Impermeable liners are not generally required for land treatment units. The waste must be placed only in soil that is above the water table. Monitoring of soil just under the treatment zone as well as groundwater monitoring is required. Maintenance of proper soil pH, careful management of waste application rate, and control of surface water runoff are all key to the operation of a land treatment unit.

5.6 Management of Radioactive and Mixed Wastes

The radioactivity of a substance is measured as the number of nuclei that decay per unit time. The standard international unit of radioactivity is called a Becquerel (abbreviated Bq), which is equal to one disintegration per second (dps). Radioactivity is also measured in Curies, a unit based on the number of disintegrations per second in one gram of radium-226 (37 billion). Note that one Curie equals 37 billion Bq.

The equation relating radioactive half-life (represented by the Greek letter tau) and the number of radioactive atoms present is

$$N = N_0 \exp\left(-\frac{0.693t}{\tau}\right)$$

In this equation, N is the number of radioactive atoms after time t and N_0 is the number of atoms at $t = 0$.

Mixed wastes are wastes that would be classified as hazardous even if they were not radioactive. These wastes can fall under the jurisdiction of more than one regulatory body, and their management is complicated by sometimes conflicting regulations.

Radioactive wastes are sometimes incinerated to reduce the volume of waste. Incineration does not destroy radioactivity, and the resulting ash is radioactive. Regulatory limits on the air emissions of radioactive materials from incineration of mixed waste have been set. Regulations allow shallow burial of ash from the incineration of radioactive waste.

Radioactive wastes are generally classified by their point of origin. High-level waste includes spent nuclear fuel and fuel reprocessing wastes. Transuranic waste is waste that contains elements with atomic numbers higher than 92, which is the atomic number of uranium. Wastes that are not high level are classified as low-level wastes, although they may be highly radioactive.

Radiological waste characterization involves determining the state of the waste and quantifying individual radionuclides in the waste. Radiography and sonar can be used to determine the physical state of the waste. Determination of individual radionuclides can be done using a variety of techniques, depending on the waste form, radionuclides involved, and level of detail/accuracy required. Measuring the radiation dose rate provides an indication of the total quantity of gamma-emitting radionuclides in a waste package, but does not identify individual radionuclides or their concentrations. Gamma spectroscopy can be used to identify the individual radionuclides and their quantities. Other techniques, such as active/passive neutron interrogation, alpha spectroscopy, and liquid scintillation counting are used for other classes of radionuclides.

Non-destructive methods of characterization are preferred because they do not involve opening a waste package to take samples. As with hazardous waste, characterization of radiological waste can be inferred from process knowledge. For example, a medical researcher who only uses a few particular radionuclides under controlled experimental conditions can use this knowledge to determine which radionuclides are present in the waste.

Storage for decay is a cost-effective way to manage short-lived, low-level radioactive wastes. For the short-lived radioisotopes typically used in medicine and research, the storage period for complete decay may be only a few weeks to a few months. After this time, the waste is no longer radioactive and can be disposed of as non-radioactive waste.

Exposure to radiation causes cancer. Health physics is the science of radiation protection. It includes protection from radiation, monitoring for the effects of radiation, and tracking the doses of radiation to which people have been exposed. The three factors for the determination of exposure to radiation are time, shielding, and distance. More information about radiation protection can be found in Sect. 6.9.

Chapter 6
Environmental Assessments and Emergency Response

6.1 Site Assessment

The National Environmental Policy Act (NEPA) requires analysis and a detailed statement of the environmental impact of any proposed federal action that significantly affects the quality of the environment. This act also applies to local (state, county, city, or industrial) projects that require a federal permit or receive funding from a federal agency. The federal government is required to use all practicable means and measures to avoid environmental degradation; preserve historic, cultural, and natural resources; and "promote the widest range of beneficial uses of the environment without undesirable and unintentional consequences."

There are three levels of NEPA analysis: (1) categorical exclusion (CATEX), (2) environmental assessment (EA), and (3) environmental impact statement (EIS). A CATEX is granted when an EA and EIS are not necessary in order to determine that no significant impacts are expected. An EA is conducted in order to determine if an EIS is necessary. If an EA indicates that no significant impacts are expected, then a finding of no significant impact (FONSI) is prepared. Otherwise, an EIS must be prepared. EISes must describe impacts as well as alternatives to the proposed action, and can be quite costly to prepare.

Environmental site assessments, which evaluate the environmental conditions at the site, are used during the preparation of environmental assessments and EISes. Site assessments involve collecting and analyzing samples, if necessary, and suggesting remediation strategies if remediation is necessary. A description of available remedial options should be included, along with their costs.

Preparation of an EA or EIS may require studies on the proposed activity's impact on environmental justice. Environmental factors such as floodplain delineations, wetland delineations, air quality, water quality, and endangered species may need to be studied. The potential destruction of natural resources such as farmland must be considered.

A. Naimpally and K. S. Rosselot, *Environmental Engineering: Review for the Professional Engineering Examination*, DOI: 10.1007/978-0-387-49930-7_6, © Springer Science+Business Media New York 2013

6.2 Hydrogeology

Wetlands can be classified by their relationship to the surrounding water table. A discharge wetland has a water surface that is lower than the water table of its surroundings. A spring or seep is a wetland that forms where the water table intersects with the soil surface, typically at the bottom of a steep slope. A throughflow wetland has a water surface that is lower than the surrounding water table on one side and higher than the surrounding water table on the other side. A recharge wetland has a water surface that is higher than the surrounding water table. In contrast to these classes of wetlands, a perched wetland is one that does not exchange water with the groundwater; its surface is above the local water table.

Soil above the water table is said to be in the unsaturated zone. Below the water table, in the saturation zone, all voids are filled with water under hydrostatic pressure. This water is known as groundwater.

An aquifer is a geologic unit that can store and transmit water. If the water table of an aquifer fluctuates in response to the water being withdrawn from or added to the aquifer, it is called an unconfined aquifer. The gauge pressure at the top of an unconfined aquifer is zero. A confined aquifer is one where the aquifer is trapped between impermeable layers. The gauge pressure of water at the top of an unconfined aquifer is not zero. An artesian aquifer is a confined aquifer whose gauge pressure is greater than zero.

Bernoulli's equation describes the driving forces that move groundwater in the saturation zone. It gives hydraulic head as

$$h = z + \frac{p}{\rho g} + \frac{v^2}{2g}$$

In this equation, z is elevation, p is groundwater pressure, ρ is groundwater density, v is the velocity of the groundwater, and g is the acceleration due to gravity.

Pressure head is the first of these terms, or

$$h_p = \frac{p}{\rho g}$$

Often, the velocity term is negligible and the equation for hydraulic head simplifies to

$$h = z + h_p$$

Darcy's equation can be used to calculate the flow rate of groundwater. It is

$$Q = -KA \frac{dh}{dx}$$

In this equation, h is hydraulic head, K is hydraulic conductivity, A is the cross-sectional area of flow, and x is the distance along the direction of flow. The term dh/dx is called the hydraulic gradient and is sometimes assigned the variable i.

The intrinsic permeability of a solid is a function of the shape and diameter of the pore spaces it contains. Intrinsic permeability and hydraulic conductivity of an aquifer are related, as follows:

$$K = \frac{k\rho g}{\mu}$$

In this equation, k is the intrinsic permeability of the soil, ρ is the groundwater density, g is the acceleration of gravity, and μ is the dynamic viscosity of the groundwater. A material with a hydraulic conductivity of 1×10^{-9} m/s or less is essentially impermeable.

Permeameters are used to measure hydraulic conductivity.

The transmissivity of an aquifer is the amount of water that can be transmitted horizontally through a unit width by the fully saturated thickness of the aquifer under a hydraulic gradient of one. It is given by

$$T = Kb$$

The variable b in this equation is the saturated thickness of the aquifer.

The potential effect of proposed actions on groundwater quality is important because groundwater provides much of the nation's drinking water. Groundwater contamination can be caused by many activities. An example of a source of groundwater contamination is a landfill whose leachate reaches groundwater. Steps are now taken to prevent landfill leachate from leaving the landfill, such as installing landfill liners and leachate collection systems. However, a liner can fail or excessive rainfall can cause leachate to spill over the edges of the liner. Another example of a source of groundwater contamination is leaking underground storage tanks. These tanks, if they leak, generally leak at a slow rate, but over a long period of time even a small leak can cause serious groundwater contamination.

When a substance is released to land, the potential for it to make its way into groundwater depends on a number of factors. One important factor is the type of soil the material has to travel through in order to reach the groundwater. Soils with the highest permeability, such as glacial outwash, are many orders of magnitude more permeable than soils with the lowest permeability, such as clays. Another factor that is important when determining the potential for a substance released to land to contaminate groundwater is the distance it has to travel before it reaches groundwater.

Contaminants that reach groundwater can migrate as they are carried by groundwater. However, they generally move slower than groundwater as they adsorb onto and desorb from the solid matrix that the groundwater is traveling through. The magnitude and distribution of contaminants in ground water is influenced mainly by (1) the timing of the release into groundwater, (2) the rate of release into groundwater, (3) groundwater flow rate and direction, (4) sorption of chemicals to solids, (5) dispersion, or the mixing or dilution of the plume, and (6) biodegradation of the contaminant.

The retardation factor of a contaminant in an aquifer is used to describe the ratio of the contaminant's velocity in groundwater relative to the groundwater. If v_x is the average linear groundwater velocity, the velocity of the contaminant is

$$v_c = \frac{v_x}{R}$$

The retardation factor, R, is calculated as follows.

$$R = 1 + \frac{\rho K_d}{\eta}$$

In this equation, ρ is the bulk density of the sediment or soil, η is the volumetric moisture content fraction in the sediment or soil, and K_d is the partition coefficient of the contaminant between the soil phase and the water phase (also called K_{sw}).

The storativity or storage coefficient of an aquifer is the volume of water taken into or released from storage per unit of surface area per unit change in piezo-metric head. It can be calculated using the following equation.

$$S = \frac{\eta \gamma_w bg}{E_w B}$$

In this equation, η is porosity, γ_w is the density of water, b is the aquifer thickness, E_w is the bulk modulus of compressibility of water (2.07×10^9 N/m^2), g is the gravitational constant, and B is the barometric efficiency of the aquifer. The storage coefficient of a confined aquifer can also be written as

$$S = \beta \gamma_w bg$$

In this equation, β is the aquifer's compressibility and b and γ_w are as before. Unconfined aquifers have storativity greater than 0.01.

6.3 Historical Considerations and Land Use Practices

Special studies of impacts on cultural resources, historical value, and social impacts of a proposed action are sometimes necessary during the preparation of an EIS.

The National Historic Preservation Act (NHPA) requires federal agencies to consider the effects of proposed actions on historic properties. The effects of a proposed action on historic buildings and archeological sites must be avoided or mitigated. This law can potentially result in a requirement for full archaeological excavation and capture at the site of a proposed action.

Measures are currently being taken to slow the loss of farmland through development.

6.4 Fate, Partitioning, and Transport of Pollutants

The concept of half-life is often used to describe the rate at which a compound leaves the area to which it was released, either because it undergoes chemical reactions that convert it to another chemical species or because it enters other environmental compartments. Half-life is related to a first-order removal rate according to the equation

$$\tau = \frac{\ln(2)}{k}$$

In this equation, τ is half-life and k is the first-order removal rate. Most environmental transformation and transport processes are best described by first-order kinetics.

A chemical's octanol-water partition coefficient is the ratio of a chemical's concentration in octanol to its concentration in the water phase of a two-phase octanol water system at equilibrium. The equation is

$$K_{ow} = \frac{C_o}{C_w}$$

In this equation, C_o is the concentration of the chemical in the octanol phase and C_w is the concentration of the chemical in the water phase. Octanol–water partition coefficients are relatively easy to measure and are sometimes used to estimate environmental partition coefficients.

The soil–water partition coefficient describes the equilibrium concentrations of a chemical that is found where soil and water contact each other. It is given by

$$K_{sw} = \frac{X}{C}$$

In this equation, X is the concentration of the chemical in the soil and C is the concentration of chemical in the water. Soil is a combination of organic matter, inorganic matter, and water, and sometimes the partitioning of a chemical between soil and water is given in terms of the organic content of the soil. This is expressed by the organic carbon partition coefficient, which is given by

$$K_{oc} = \frac{C_{soil}}{C_{water}}$$

In this equation, C_{soil} is the concentration of the chemical in the organic carbon portion of the soil and C_{water} is the concentration of the chemical in the water phase. The relationship between the soil–water partition coefficient and the organic carbon partition coefficient is

$$K_{sw} = K_{oc} f_{oc}$$

In this equation, f_{oc} is the fraction of organic compound in the soil.

Pollutants can contaminate living things when they diffuse through flesh or are ingested or inhaled. Living tissue sometimes has a higher concentration of a pollutant than the surrounding environment. Bioconcentration factors are used to describe this effect. A chemical's bioconcentration factor is

$$BCF = \frac{C_{org}}{C}$$

In this equation, BCF is bioconcentration factor, C_{org} is the concentration of the pollutant in the organism, and C is the concentration of pollutant in the environment. Each species of organism has a different bioconcentration factor for each pollutant.

Deposition is a means of pollutant transport from particles in air to soil or from particles in water to sediment. Stoke's Law can be used to determine the terminal settling velocity of particles in still air or water. It is

$$v_t = \frac{g(\rho_p - \rho_f)d^2}{18\mu}$$

6.5 Industrial Hygiene, Health, and Safety

Many compounds found in the workplace are regulated under the Occupational Safety and Health Act (OSHA). OSHA regulations apply strictly to employees. OSHA-regulated chemicals have been assigned threshold air concentrations that cannot be surpassed in the workplace. Permissible Exposure Limits, or PELs, give the maximum airborne concentration of a contaminant to which an employee may be exposed over a given period of time. Recommended Exposure Limits, or RELs, are expressed as 8- or 10-h time weighted averages for a 40-h workweek and/or maximum levels for time periods ranging from instantaneous to 120 min.

Regulatory limits are generally based on values taken from dose–response curves. These curves are constructed from human studies where possible. Most often, the curves reflect toxicity studies conducted on animals.

The rate of vaporization from a spilled liquid can be estimated using the following equation.

$$Q_{vapor} = \frac{K(MW)A_{liquid}P_{sat}}{RT_{liquid}}$$

In this equation, K is the mass transfer coefficient, MW is the molecular weight of the vaporized compound, A_{liquid} is the surface area of the spill, P_{sat} is the saturation vapor pressure of the vaporized compound, R is the ideal gas constant, and T_{liquid} is the absolute temperature of the spilled liquid. In a ventilated space, the rate of vaporization of a spilled liquid can be used to calculate the concentration of the vapor in the air, as follows.

$$C = \frac{Q_{vapor} R T_{air}}{k Q_{vent} P (MW)}$$

In this equation, T_{air} is the absolute temperature of the air, k is the nonideal mixing factor, Q_{vent} is the ventilation rate, and P is the absolute ambient pressure.

A vapor-air mixture is capable of igniting and burning if the vapor is present at concentrations above its lower flammability limit and below its upper flammability limit.

Sometimes, a safety hazard is created when two substances are combined. Tables that show the consequences of mixing various substances are available. One such table is given in Table 6.1. To use this table, determine if a consequence code is given in the square where the row of one of the substances to be mixed meets the column of the other substance. If no adverse consequence is expected, the square will contain no consequence code.

6.6 Security, Emergency Plans, and Incident Response Procedures

The Secretary of Homeland Security is required to develop and administer a National Incident Management System (NIMS). NIMS provides a consistent nationwide template to enable all government, private-sector, and nongovernmental organizations to work together during domestic incidents. The intent of NIMS is to have a system in place that is applicable across a full spectrum of potential incidents and hazard scenarios, regardless of size or complexity. NIMS is also intended to improve coordination and cooperation between public and private entities in a variety of domestic incident management activities. Federal departments and agencies are required to make the adoption of NIMS by state and local organizations a condition for federal preparedness assistance.

Jurisdictions can comply with NIMS by adopting the Incident Command System (ICS). ICS is the standard management structure for responses to domestic incidents. ICS requires the use of common terminology for facilities and positions within the response team. This helps clarify the activities that take place at a specific facility, and identifies what members of the organization can be found there. For example, the Incident Commander can be found at the Incident Command Post. Only the Incident Commander is called Commander—and there is only one Incident Commander per incident. Only the heads of Sections are called Chiefs. Learning and using standard terminology helps reduce confusion between the day-to-day position occupied by an individual and his or her position at the incident.

Table 6.1 Compound compatibility

KEY:

REACTIVITY CODE	CONSEQUENCES
H	HEAT GENERATION
F	FIRE
G	INNOCUOUS AND NON-FLAMMABLE GAS
GT	TOXIC GAS GENERATION
GF	FLAMMABLE GAS GENERATION
E	EXPLOSION
P	POLYMERIZATION
S	SOLUBILIZATION OF TOXIC MATERIAL
U	MAY BE HAZAROUS BUT UNKNOWN

EXAMPLE:

H	HEAT GENERATION, FIRE,
F	AND TOXIC GAS
GT	GENERATION

Reactivity Group

No.	Name	1	2	3	4	5	6	7	8	9	10	11	12	13	14	15	16	17	18
1	Acid, Minerals, Non-Oxidizing	1																	
2	Acids, Minerals, Oxidizing		2																
3	Acids, Organic	G H	H	3															
4	Alcohols & Glycols	H	H P	H P	4														
5	Aldehydes	H P	H G P	H P	H G	5													
6	Amides	H	GT				6												
7	Amines, Aliphatic & Aromatic	H	GT	H		H		7											
8	Azo Compounds, Diazo Comp, Hydrazines	H G	H GT G	H G	H G	H		H	8										
9	Carbamates	H G	H GT						H G	9									
10	Caustics	H	H	H		H			G	H G	10								
11	Cyanides	GT GF	GT GF	GT GF				U	G			11							
12	Dithiocarbamates	GF F	H GF F	GF GT		GF GT			H G				12						
13	Esters	H	H F	H					H G		H			13					
14	Ethers	H	H F						H G						14				
15	Fluorides, Inorganic	GT	GT													15			
16	Hydrocarbons, Aromatic	H	H F														16		
17	Halogenated Organics	H GT	GT	H F				H GT G			H		H GF					17	
18	Isocyanates	H G	H F GT	H P	H P		H P		H P G U		H U G								18

(continued)

Table 6.1 (continued)

	1	2	3	4	5	6	7	8	9	10	11	12	13	14	15	16	17	18	19	20	21	104	105	106	107
19 Ketones	H	H F	H F	GF H F	GF H F	GF H	GF H	H G	GF H	GF H	GF H	GF GT H	GF H			H F	H	H	19						
20 Mercaptans & Other Organic Sulfides	GT GF	H F	GF GT	GF H F	GF H F	GF H	F GT H	H G	GF H	GF H	GF H	GF GT H	GF H			H	H GT	GF H	GF H	20					
21 Metal, Alkali & Alkaline Earth, Elemental	GF H F	GF H F	GF H F	GF H F	GF H F	H F	GT H	GF H	F GT	GF H	GF H H	GF GT H	GF H			H E	H	GF H	GF H	GF H	21				
104 Oxidizing Agents, Strong	H GT		H GT	H GT	H F	H F	H F GT	H E	H F GT		H E GT	H F GT	H F	H F		H F	H GT	H GT	H F F	H F GT	H F GE	104			
105 Reducing Agents, Strong	H GF	H F GT	H F GT	H GF F	GF H F	GF H	GF H	H G				H GT H	H GT F				H T	H G	GF H	GF H	GF H		105		
106 Water & Mixtures Containing Water	H	H						G										H G	H		GF H			106	
107 Water Reactive Substances	EXTREMELY REACTIVE! Do Not Mix With Any Chemical Or Waste Material																					104	105	106	107
	1	2	3	4	5	6	7	8	9	10	11	12	13	14	15	16	17	18	19	20	21	104	105	106	107

Incident management prepares incident action plans, usually based on 12-h periods, that are used to issue assignments, plans, procedures, and protocols. Resources, including all personnel, facilities, and major equipment and supply items used to support incident management activities are assigned common designations. Resources are "typed" with respect to capability to help avoid confusion and enhance interoperability.

ICS also requires integrated communications, an orderly chain of command, check-in for all responders regardless of agency affiliation, and unity of command (each individual involved in incident operations is assigned only one supervisor). Maintaining adequate span of control is required in ICS; a ratio of one supervisor to five reporting elements is recommended.

A facility can do a risk assessment in order to help determine an appropriate amount of money to spend on risk reduction. These assessments can be quantitative, but even when they are quantitative, many of the values used are subjective and highly uncertain. Inclusion of the time value of money and tax rates in these calculations would be questionable. This discussion will assume risk to an asset that does not involve human safety.

An asset's single loss expectancy (SLE) is the amount that would be lost from a single occurrence of a risk. This can be estimated by determining how much the asset would cost to replace or repair and how much revenue would be lost before the asset can be restored. The value of advertising that would be required to overcome negative public opinion due to loss or disruption of the asset can also be included in this estimate. The annual rate of occurrence (ARO) of a risk is the number of times the risk can reasonably be expected to occur each year. This is particularly difficult to evaluate. Sometimes it can be determined from previous records. Annual loss expectancy (ALE) is

$$ALE = ARO \times SLE$$

ALE can be used to budget mitigating factors for risk reduction. The return on security investment, or ROSI, is

$$ROSI = ALE_0 - ALE_{final} - \text{annual cost of control}$$

6.7 Fundamentals of Epidemiology and Toxicology

The dose–response relationship is a fundamental and essential concept in toxicology. It correlates exposures and the spectrum of induced effects. Generally, more severe responses are associated with higher doses. Dose–response relationships are based on observed data from experimental animal, human clinical, or cell studies. The dose–response curve normally takes the form of a sigmoid curve. In most cases, no effect is observed at small doses. The point at which toxicity first appears is known as the threshold dose level. From that point, the curve increases with higher dose levels until it reaches an asymptote where 100 % of the population is affected.

Within a population, a wide variety of responses to the same dose of toxicant may be encountered, with some individuals exhibiting greater susceptibility to the toxicant and others exhibiting greater resistance. A graph of the individual responses can usually be depicted as a bell-shaped standard distribution curve. A large standard deviation indicates greater variability in response.

There are two types of health effects: chronic and acute. Acute health effects are those that would be felt upon a one-time exposure to a chemical and can be temporary or permanent, and mild or severe. One measure of a chemical's acute toxicity is its LD_{50}, which is the dose of chemical required to kill half of the population in a study group. A compound with a high LD_{50} is less toxic than a compound with a low LD_{50}.

Chronic health effects are those that occur after a long-term exposure to a chemical. Chronic health effects can be divided into two groups: cancer and non-cancer. Substances that cause cancer are called carcinogens. Non-cancer health effects can have diverse manifestations including retardation, infertility, and abnormal growth rates.

There are four common types of epidemiological studies. A cohort study follows a group of healthy people with different levels of exposure and assesses what happens to their health over time. Cohort studies are expensive, time-consuming and the most logistically difficult of all the studies. Because of the number of people who must be studied, cohort studies can only be used for relatively common diseases. Their advantage is that exposure precedes the outcome, but the level of exposure must be able to be characterized and it must be rare enough that effects can be tied to the exposure and not to other variables.

Case–control studies examine the prior exposure of individuals with a health condition compared to individuals without the health condition. Case–control studies allow rare health conditions to be studied without following thousands of people. There is a potential for bias because subjects are chosen based on outcome.

In occupational studies, working people with particular jobs or exposures are selected as subjects. Workers often have substantially higher exposures to certain risk factors than the typical population, so the chances of detecting an exposure effect are increased. However, workers differ substantially from one another in terms of risks, and the working population has many social and economic differences from the nonworking population. For example, young children and the very elderly are not generally exposed to occupational conditions. Another complication of occupational studies is that workers may be exposed to more than one risk factor.

In a cross-sectional study, the current health of individuals and their current level of exposure are assessed. This type of study does not accommodate health conditions that require time to develop. Also, current health conditions are not necessarily reflective of current exposures.

Epidemiologists use statistics to determine if the effects they are seeing are real or are due to chance fluctuations.

The rate of disease is defined as the number of new diagnoses per number of people per unit time. Subtracting the rate of disease in a population without the risk factor from the rate of disease in an exposed population provides a value for the attributable risk due to the risk factor. The attributable risk is useful because it allows for the calculation of the extra cases of disease expected in a specific population over a year, given exposure to the risk factor of concern. Rate ratio, or relative risk, is the ratio of disease in the exposed population to disease in the unexposed population.

6.8 Exposure Assessment and Risk Characterization

Information about the dose of chemical an individual receives and the degree of toxicity of that chemical are needed in order to assess whether the individual is being exposed to harmful levels of the chemical. People are exposed to chemicals via several routes, including ingestion by mouth, inhalation, and absorbtion through the skin. Values used to calculate the exposure to a chemical are given in Table 6.2 for adults and children. Once a body is exposed to a chemical, an absorption factor specific to the chemical and the route of exposure is multiplied by the total exposure in order to determine the dose that enters body tissue.

A substance's oral slope factor is equal to the risk of getting cancer from ingesting a given quantity of that substance per body weight per unit of time. The risk of getting cancer due to ingestion of a chemical is

$$\text{risk}_{\text{ing}} = \frac{C_{\text{ing}}I \times \text{oral slope factor}}{\text{BW}}$$

In this equation, C_{ing} is the concentration of the chemical in the material that is ingested, I is the intake rate of the ingested material, and BW is the individual's body weight.

The drinking water unit risk is the risk of getting cancer per mass of chemical per volume of water. The risk of getting cancer due to a lifetime of drinking water containing a chemical at concentration C_{water} is

$$\text{risk}_{\text{water}} = C_{\text{water}} \times \text{drinking water unit risk}$$

A chemical's air unit risk from breathing air containing a carcinogen is the risk of getting cancer per mass of chemical per volume of air. The risk of getting cancer due to a lifetime spent inhaling a chemical at concentration C_{air} in air is

$$\text{risk}_{\text{air}} = C_{\text{air}} \times \text{air unit risk}$$

Notice that an average intake rate for water and an average inhalation rate for air are assumed when developing unit risks, as opposed to the oral slope factor. The ability of a chemical to cause cancer increases with increasing oral slope factor, air unit risk, or drinking water unit risk.

Table 6.2 Values used in calculating exposure

	Soil ingestion rate
	100 mg/d (children > 6 years old)
	200 mg/d (children 1–6 years old)
	Exposure duration
	30 years at one residence (adult)
	6 years (child)
	Body mass
	70 kg (adult)
	10 kg (child)
	Averaging period
	Non-carcinogens, actual exposure duration
	Carcinogens, 70 years
	Water consumption rate
	2.0 L/d (adult)
	1.0 L/d (child)
	Inhalation rate
	0.83 m^3/h (adult)
	0.46 m^3/h (child)

The risks for all exposure routes (ingestion, breathing, drinking, etc.) are summed to get the total risk for a chemical. Usually a risk greater than 10^{-6} (one in a million) or 10^{-4} (one in ten thousand) is considered unacceptable.

The no observable adverse effect level (NOAEL) dose is one measure of the potential for a chemical to cause chronic non-cancerous health effects in a particular species of animal. No harmful effects are expected at or below a substance's NOAEL. For ingested chemicals, the reference dose, or RfD, is the NOAEL divided by a safety factor. This safety factor is intended to conservatively correct for uncertainty in applying NOAELs for animals to humans. It is also intended to correct for variations in sensitivity among individuals and other uncertainties. The lower a chemical's RfD is, the more potent is its ability to cause chronic health effects. The hazard quotient due to ingestion of a chemical whose concentration is C_{ing} is

$$HQ_{ing} = \frac{C_{ing}I}{RfD \times BW}$$

Reference concentrations are given for air inhalation (RfC_{air}) and water intake (RfC_{water}). These concentrations assume a standard inhalation and water consumption rate. The hazard quotient for inhalation of a chemical at concentration C_{air} is given by

$$HQ_{air} = \frac{C_{air}}{RfC_{air}}$$

The hazard quotient for drinking water that has a chemical concentration of C_{water} is given by

$$HQ_{water} = \frac{C_{water}}{RfC_{water}}$$

A chemical's hazard index (HI) is the sum of the hazard quotients for all exposure routes. A hazard index greater than one is generally considered to be unacceptable.

6.9 Radiation Protection (Health Physics)

Health physics is the science of radiation protection. Information on measuring radioactivity is presented in the earlier section on management of radioactive and mixed wastes.

Most of the tissue damage caused by radiation is due to ionization of water and other molecules. There are four types of ionizing radiation: alpha particles, beta particles, gamma rays, and neutron radiation.

An alpha particle consists of a helium nucleus and contains 2 neutrons and 2 protons. Alpha particles cannot penetrate skin, aluminum, lead, or concrete. They can, however, enter the lungs via inhalation and cause tissue damage. Beta particles are made up of electrons or positrons, which cannot penetrate aluminum, lead, or concrete but are capable of penetrating skin. Gamma rays are short-wave electromagnetic radiation, and can penetrate skin and aluminum, but not lead or concrete. Low energy gamma rays are used in X-rays. Neutron radiation is produced during spontaneous fission in nuclear reactors and can penetrate skin, aluminum, and lead, but not concrete.

For any given radioactive source, three factors affect the dose of radiation an individual receives: the distance from the source, the time over which the individual is exposed, and any shielding between the source and the individual.

Radiation doses are measured as energy imparted per unit of tissue mass. The standard unit of measurement for radiation doses is the Gray, which is equal to 1 J/kg. Dose is also measured in rads. One rad is equal to 0.01 Gray.

The amount of time an individual is exposed to radiation is directly proportional to the dose of radiation received.

The dose of radiation from a point source of gamma and X-rays reduces with distance as follows:

$$\text{dose} \propto \frac{1}{r^2}$$

In this equation, r is distance from the source. Another way of expressing this is

$$\text{flux at point 2} = \text{flux at point 1} \times \left(\frac{r_1}{r_2}\right)^2$$

In this equation, r_1 is the distance from point 1 to the source and r_2 is the distance from point 2 to the source.

The deposition of energy is only one aspect of the potential of radiation to cause biological damage. The damage caused per unit of deposited energy is greater when it is deposited over a shorter distance. Therefore, alpha particles, which deposit their energy over a very short distance, cause more damage per unit of energy than gamma rays, which deposit their energy over a longer distance. The density of biological matter in which the energy is deposited is also important, and different organs have varying sensitivity to radiation. The concept of relative biological effectiveness (RBE) has been created to try to capture the relative efficiency of various kinds of radiation in producing biological damage.

6.10 Vector Control and Sanitation, Including Biohazards

A vector is an animal that does not cause disease, but that spreads disease by carrying pathogens from one host to another. Mosquitoes, for example, serve as vectors for the disease caused by West Nile Virus. They ingest the virus when they bite an infected animal and regurgitate some of the virus into their next host. Rats served as the vector for the infamous Black Death in Europe during the Middle Ages. The fleas that infected humans with bubonic plague were carried to Europe by infected rats. Some diseases, such as hantavirus illnesses, can be transmitted directly from rodents to humans as well.

Public health management requires control of disease-carrying pests such as rats, mice, and mosquitoes. While poisoning and to a lesser extent trapping of rats and mice and spraying to kill adult mosquitoes can achieve a temporary reduction in pest populations, the most effective programs integrate better housekeeping and maintenance practices with use of pesticides and introduction or husbandry of predators. This approach is called integrated pest management (IPM).

Control of mice and rats requires the elimination of food and shelter to whatever extent is possible. Trash and food must be stored in containers that are inaccessible to rodents. Debris that provides optimal nesting conditions must be removed. Proper community sanitation so that trash containers are not allowed to overflow is important. Structures should be rodent-proofed and landscaping should be designed so that foliage does not overhang buildings.

Control of mosquitoes relies on removal of standing water that is required for the larval stage of the mosquito life cycle. Old tires and plugged rain gutters are sources of standing water that can easily be remedied. Where standing water cannot be removed, mosquito-eating fish or applications of mosquito larvicide might be introduced.

Biohazards include cultures of pathogens, human blood and tissues, infected animals, and any equipment that comes in contact with a biohazard.

Risk group 1 biohazard agents are those that do not cause disease in humans or that do not cause disease in healthy adults. Risk group 2 agents are those that cause disease in humans, but not serious disease. Examples of risk group 2 agents include *Escherichia coli* (often the culprit in foodborne illnesses), adenovirus (a cold virus), and herpes simplex virus. Risk group 3 agents are associated with a lethal or serious disease, and include HIV and *Mycobacterium tuberculosis*. Risk group 4 agents cause lethal or serious diseases that cannot typically be treated, and include the ebola virus.

Safety practice levels that equate to the risk group levels are required. For example, personal protective equipment must be worn when using agents from risk group 2 or higher. Also, when biohazard agents of risk group 2 or higher are present, a biohazard placard should be placed on the door where it can be easily seen. Medical sharps present the greatest risk of getting a bloodborne pathogen infection on the job and must be disposed of immediately in approved containers.

Some biohazard agents present a special concern because they can transmit to humans via an aerosol route. These agents are much more difficult to contain than others.

6.11 Noise Pollution

Noise pollution can cause hearing loss, which affects quality of life. The impact of a given noise can be predicted based on its sound pressure level, which is given in decibels (dB):

$$SPL = 10\log_{10}\left(\frac{P^2}{P_0^2}\right)$$

In this equation, P is the measured sound pressure from the source and P_0 is the reference sound pressure (which is 2×10^{-5} Pa).

The weakest sound the average human ear can hear is 0 dB, and the loudest tone possible is 194 dB. The sound pressure level of normal conversation is 60 dB, of a ringing telephone is 80 dB, of an ambulance siren is 120 dB, and of a jet engine at takeoff is 140 dB.

When there is more than one source of sound, the total sound pressure level is given by

$$SPL_{TOTAL} = 10\log_{10}\sum 10^{SPL/10}$$

Point source attenuation occurs when the sound pressure level changes at different distances from a point source. It can be expressed in decibels in the following equation:

$$\Delta SPL = 10\log_{10}\left(\frac{r_1}{r_2}\right)^2$$

In this equation, r_1 and r_2 are the distances from the source of sound.

Line source attenuation is the change in sound pressure level from a line source, and is expressed in decibels as follows:

$$\Delta\text{SPL} = 10\log_{10}\left(\frac{r_1}{r_2}\right)$$

OSHA regulates noise pollution in the workplace. Permissible noise exposure is based on both the attenuated time-averaged sound pressure level of the noise and the length of time of exposure. Attenuation of a noise pollution source can be based on distance from the source or on ear protection. Over an 8-h day, the average sound pressure level is permitted to be 90 dBA. Over a 2-h period, the average sound level may not exceed 100 dBA, and over a ¼-h or less time period, the average sound level may not exceed 115 dBA.

6.12 Indoor Air Quality

Indoor air can have higher concentrations of harmful pollutants than outdoor air in even the most polluted cities. Also, most people spend the majority of their time indoors. Sources of indoor air pollution include tobacco smoke, formaldehyde emissions from building materials, paint fumes, mold, household cleaning products, and pesticides used in the home. Radon, a carcinogenic radioactive gas, can enter homes under certain conditions and is estimated to be the second leading cause of lung cancer in the United States. Motor vehicle fumes can enter a home via an attached garage. Carbon monoxide is released from appliances that burn natural gas. Of all of these pollutants, radon and tobacco smoke present the highest level of concern for public health.

Human health effects from indoor air pollution include headaches, tiredness, dizziness, nausea, and throat irritation. More serious effects include cancer and exacerbation of chronic respiratory diseases, such as asthma. In addition to being a carcinogen, tobacco smoke causes eye, nose and throat irritation. The development of asthma in children is associated with poor indoor air quality.

Chapter 7
Error and Uncertainty in Environmental Measurements

One of the key issues in environmental monitoring and sampling is the degree of error in the measurements and the uncertainty in results taken from measured values.

Two factors affecting the error in a measurement are the instrument's precision and its accuracy. Precision is related to the number of significant digits or decimal points that can be read from an instrument. Accuracy describes how close the instrument's readings are to the true value of the property being measured.

The Kline–McClintock equation can be used to estimate the uncertainty in the value of a function that depends on measured variables. This equation is

$$w_R = \sqrt{\left(w_1 \frac{\partial f}{\partial x_1}\right)^2 + \left(w_2 \frac{\partial f}{\partial x_2}\right)^2 + \cdots + \left(w_n \frac{\partial f}{\partial x_n}\right)^2}$$

In this equation, R is a function of x_1, x_2 ... x_n, and w_1, w_2 ... w_n are the uncertainties in the variables x_1, x_2 ... x_n. This function gives a better estimate of the uncertainty in R than one would get by assuming that the uncertainties of the variables accumulate on a worst-case basis.

For example, let

$$R = a + b + c$$

The value of a is 3.0 ± 0.1 cm, the value of b is 2 ± 0.2 cm, and the value of c is 0.5 ± 0.1 cm. If all three measurements are simultaneously at either the extreme high or low end of their inaccuracies, then the uncertainty in R could be as high as ± 0.4 cm. The Kline–McClintock equation gives an uncertainty of

$$w_R = \sqrt{w_a^2 + w_b^2 + w_c^2}$$
$$= \sqrt{0.1 \text{ cm}^2 + 0.2 \text{ cm}^2 + 0.1 \text{ cm}^2}$$
$$= 0.2 \text{ cm}$$

This value (0.2 cm) is considered to be a better estimate of the uncertainty in R. The standard deviation of a sampling of a population is equal to

A. Naimpally and K. S. Rosselot, *Environmental Engineering: Review for the Professional Engineering Examination*, DOI: 10.1007/978-0-387-49930-7_7, © Springer Science+Business Media New York 2013

$$s = \sqrt{\frac{\sum (x - \bar{x})^2}{n - 1}}$$

In this equation, x is the measured value, $\bar{x}$ is the mean of the measured values, and n is the number of measured values. The coefficient of variation as a percent is defined as

$$CV = \frac{100s}{\bar{x}}$$

A Type I error is one that results in a false positive, such as detecting the presence of a chemical when the chemical is not actually present. The probability of making a Type I error is assigned the variable α.

A Type II error is one that results in a false negative, such as failing to detect the presence of a chemical when the chemical actually is present. The probability of making a Type II error is assigned the variable β.

Detection limits are an important quality of environmental monitoring instruments. Examples of detection limit definitions are described below to illustrate the ways in which detection limits might be expressed. These examples focus on instrument limitations, not on sampling errors. Note that definitions of various detection limits are not universal.

The *instrumental detection level* is the concentration of a chemical that produces a signal that is five times greater than the instrument's signal to noise ratio.

The *lower level of detection* might be defined as the concentration of chemical that produces a signal that is about $2(1.645\ s)$ above the mean of blank analyses, where s is the standard deviation of the blank analyses. At this definition of the lower level of detection, both α and β are 5 %.

The *method detection level* might be defined as the concentration of chemical at which β is 1 %. At this concentration, the mean of seven replicate samples would be $3.14\ s$ above the blank signal, where s is the standard deviation of the seven replicates.

The *level of quantification* is the concentration of chemical that produces a signal that can be detected within specified levels. This may be defined as the concentration that produces a signal that is $10\ s$ above the blank signal, where s is the standard deviation of blank analyses.

A sampling and analysis plan's confidence level is defined as the probability of not making a Type I error, or

$$\text{confidence level} = 1 - \alpha$$

In other words, the confidence level is the probability of not having a false positive result. Power is defined as the probability of not making a Type II error, or

$$\text{power} = 1 - \beta$$

Power is the probability of not having a false-negative result.

The *minimum detectable relative difference* between background concentrations of a chemical and that chemical's concentration in contaminated material is the relative increase over background that is detectable with a probability of $1 - \beta$. The relative increase over background is given by

$$\text{relative increase over background} = \frac{100(\mu_s - \mu_B)}{\mu_B}$$

In this equation, μ_s is the contaminated sample concentration and μ_B is the background concentration.

Exam 1

1. At 20 °C, the Henry's Law constant for benzene is 240 atm. The concentration of benzene in a body of water at this temperature is 3×10^{-5} mol/L. The concentration of benzene in the air above the water is most nearly

   (a) 0.00013 ppm
   (b) 0.0072 ppm
   (c) 130 ppm
   (d) 7200 ppm

   For problems 2 and 3, a population of 300 air samples has a mean concentration of 88 ppm for a compound of interest. The standard deviation of the population is 14 ppm and the distribution curve is normal. A sample of 13 of the air samples is re-evaluated for concentration of the pollutant.

2. The standard deviation of the sample's distribution is most likely to be nearest to

   (a) 1.1
   (b) 2.0
   (c) 14
   (d) 28

3. The probability that a sample's concentration will differ from the population mean by less than 5 ppm is most nearly

   (a) 14 %
   (b) 21 %
   (c) 28 %
   (d) 88 %

4. An unflanged hood must have a capture velocity of 1.5 m/s. The farthest contaminant release point is 0.4 m from the hood, and the area of the hood is 1,000 cm^2. The estimated flow rate through the hood is most nearly

   (a) 1.9 m^3/s
   (b) 2.6 m^3/s
   (c) 6.2 m^3/s
   (d) 1,500 m^3/s

A. Naimpally and K. S. Rosselot, *Environmental Engineering: Review for the Professional Engineering Examination*, DOI: 10.1007/978-0-387-49930-7, © Springer Science+Business Media New York 2013

5. The half-life of DDT in sediment is 4 years. The removal mechanism is first order. In an accidental spill, 4 kg of DDT are released to a shallow lake, where virtually all of the DDT immediately enters the sediment phase. After 18 months, the DDT remaining in the lake is most nearly

   (a) 0.18 kg
   (b) 2.5 kg
   (c) 2.7 kg
   (d) 3.1 kg

For problems 6 through 9, the particle size distribution of a sample taken from an exhaust system is determined from 12 particles. The diameters of the particles, which are perfect spheres, are 0.037, 0.14, 0.30, 0.31, 0.58, 0.59, 0.60, 0.61, 1.1, 1.1, 2.4, and 9.5 μm.

6. The geometric mean diameter of these particles is most nearly

   (a) 0.47 μm
   (b) 0.59 μm
   (c) 1.1 μm
   (d) 1.4 μm

7. The geometric standard deviation of the diameter of these particles is most nearly

   (a) 0.60 μm
   (b) 1.4 μm
   (c) 2.6 μm
   (d) 4.0 μm

8. The distribution of these points most nearly resembles a

   (a) log-normal distribution
   (b) normal distribution
   (c) uniform distribution
   (d) bimodal distribution

9. The percentage of particles that are $PM_{0.1}$ is most nearly

   (a) 1 %
   (b) 8 %
   (c) 92 %
   (d) 100 %

10. The nitrogen content of dry air is most nearly

   (a) 22 %
   (b) 34 %
   (c) 66 %
   (d) 78 %

11. Methylene chloride's reaction rate with hydroxyl radicals in the atmosphere is approximately 1.3×10^{-13} cm^3/molecule-s. Reaction with hydroxyl radicals is this compound's major removal mechanism from the atmosphere. The concentration of hydroxyl radicals in the atmosphere is 8.5×10^5 molecule/cm^3. Methylene chloride's half-life in the atmosphere is most nearly

    (a) 0.0095 day
    (b) 0.014 day
    (c) 72 days
    (d) 104 days

12. More photochemical smog is produced on sunny days in general because

    (a) more smog-forming pollutants are released on sunny days
    (b) some of the chemical reactions in the smog formation cycle require light
    (c) there is more atmospheric stability on sunny days
    (d) this statement is not true; whether a day is sunny or not has no impact on smog formation

13. Beta particles are stopped by

    (a) concrete, but not lead
    (b) lead and concrete, but not aluminum
    (c) aluminum, lead, and concrete, but not skin
    (d) aluminum, lead, concrete, and skin

14. A sample containing methanol, n-butane, n-hexane, and n-octane is sent through a gas chromatographer with a photoionization detector. There are four peaks in the resulting chromatograph. The peak with the second longest retention time is much larger than the other peaks. The compound in this sample with the highest concentration is

    (a) methanol
    (b) n-butane
    (c) n-hexane
    (d) n-octane

15. The combustion efficiency of a particular incinerator must not drop below 99.95 %. If the concentration (dry basis) of carbon dioxide leaving the incinerator is 140,000 ppm, the concentration of carbon monoxide must be less than

    (a) 5 ppm
    (b) 30 ppm
    (c) 55 ppm
    (d) 70 ppm

For problems 16 and 17, use the following conditions. The concentration of mercury in soil is 30 ng/g, the concentration in air is 2 ng/m^3, the concentration of mercury in fish is 1 ppm, and the concentration of mercury in drinking water is

20 ng/L. Average fish consumption is 6.4 kg/years for children 1–6 years of age and 45.8 kg/year for adults.

16. The exposure to mercury by the average 5-year-old child under these conditions is most nearly

    (a) 6.0×10^{-7} mg/kg/day
    (b) 2.0×10^{-6} mg/kg/day
    (c) 1.8×10^{-3} mg/kg/day
    (d) $2.8 \times 10^{-}$ mg/kg/day

17. The exposure to mercury by the average adult under these conditions is most nearly

    (a) 4.3×10^{-8} mg/kg/day
    (b) 5.7×10^{-7} mg/kg/day
    (c) 1.8×10^{-3} mg/kg/day
    (d) 2.8×10^{-1} mg/kg/day

18. A quarterly leak detection and repair program with a leak definition of 500 ppm to reduce fugitive emissions at a chemical manufacturing plant is expected to have an annualized cost of $350,000. The current quarterly leak detection and repair program with a leak definition of 10,000 ppm has an annualized cost of $204,000. The lower leak definition of 500 ppm will result in a reduction in fugitive emissions of 70 Mg/year. The value of the product lost in fugitive emissions is $2.10/L. Assume the density of the captured losses is 0.9 kg/L. The net effect of lowering the leak definition from 10,000 ppm to 500 ppm is most nearly

    (a) a net annualized cost of $150,000/year
    (b) a net annualized cost of $80,000/year
    (c) a net annualized cost of $17,000/year
    (d) a net annualized savings of $17,000/year

19. A hazardous waste incinerator is used to treat a waste that contains carbon tetrachloride. The flow rate of carbon tetrachloride into the incinerator is 20 kg/d. The mass flow rate of carbon tetrachloride out of the incinerator is 0.0045 kg/d. The destruction and removal efficiency is most nearly

    (a) 0.9998 %
    (b) 9.998 %
    (c) 99.75 %
    (d) 99.98 %

20. Integrated pest management (IPM) for mosquitoes

    (a) eliminates spraying to kill adult mosquitoes
    (b) may include the use of mosquito larvicides and mosquito-eating fish
    (c) includes the elimination of standing water wherever possible
    (d) (b) and (c)

21. The mean concentration of a contaminant in samples of soil at a proposed development is 240 ppm. The standard deviation of the samples is 60 ppm. If the probability of getting a false positive is 20 %, the probability of getting a false negative is 5 %, and the desired minimum detectable relative difference is 10 %, the number of samples which must be taken is most nearly

 (a) 19
 (b) 41
 (c) 69
 (d) 155

22. The concentration of carbon dioxide in the troposphere is 350 ppm. The molar concentration of carbon dioxide in air at standard temperature and pressure is most nearly

 (a) 1.0×10^{-5} mol/L
 (b) 1.6×10^{-5} mol/L
 (c) 3.5×10^{-4} mol/L
 (d) 4.5×10^{-4} mol/L

23. A facility generates a sodium hydroxide waste stream and a hazardous waste stream that contains perchloroethylene. Mixing of these wastes would result in the following consequence(s):

 (a) no consequences
 (b) the generation of heat and flammable gas
 (c) the generation of a toxic and flammable gas
 (d) polymerization

24. The air unit risk for nickel is $0.00026 \ (\mu g/m^3)^{-1}$. In a town, 20,000 people are exposed to an average concentration of nickel in air of $0.001 \ mg/m^3$ over the course of their lifetime. The number of expected cancer cases due to the presence of nickel in the air is most nearly

 (a) air unit risk cannot be used to predict cancer
 (b) <1
 (c) 5
 (d) 10

25. The activity of a mass of radioactive waste is 10 Bq. The half-life of the waste is 13 days. The time at which the activity reaches 1.0 Bq is most nearly

 (a) 1.9 days
 (b) 30 days
 (c) 37 days
 (d) 43 days

26. A waste sample is considered ignitable under the Resource Conservation and Recovery Act (RCRA) if it

(a) has a flash point of 120 °F and is an aqueous solution containing 10 % alcohol by volume

(b) is an ignitable compressed gas

(c) is a strong reducing agent

(d) is liquid that ignites when a flame is applied to it

27. A hole is punched 4 m from the top of a hazardous waste tank that is open to the atmosphere. The hole is 80 cm^2 and its discharge coefficient is 0.55. The tank is originally full of a hazardous liquid waste that has a density of 0.9 g/mL. The rate at which the hazardous waste flows out of the tank at the time of the puncture is most nearly

(a) 25 kg/s

(b) 35 kg/s

(c) 110 kg/s

(d) 310 kg/s

28. A shaker/woven baghouse is to be used to trap silica dust. The volumetric flow rate of air to be treated is 0.3 m^3/s. The necessary fabric area is most nearly

(a) 0.38 m^2

(b) 8.6 m^2

(c) 23 m^2

(d) 5,400 m^2

For problems 29 and 30, consider a packed-bed wet scrubber that is used to remove hydrogen chloride from an otherwise clean air stream. The scrubber fluid is not recycled.

29. The pH of the scrubber fluid as it enters the scrubber is 9 and the pH of the scrubber fluid as it exits the scrubber is 8. The flow rate of the scrubber fluid is 0.2 L/s. The rate of increase of hydronium ions in the scrubber fluid is

(a) 1.8×10^{-9} mol/s

(b) 2.8×10^{-9} mol/s

(c) 2.8×10^{-3} mol/s

(d) 0.20 mol/s

30. The mass flow rate of hydrogen chloride removed is most nearly

(a) 1.8×10^{-9} g/s

(b) 6.5×10^{-8} g/s

(c) 1.0×10^{-7} g/s

(d) 1.4×10^{-7} g/s

31. Commensal rodents are known as vectors because they

(a) are non-native species

(b) consume food that is meant for humans

(c) can have disease-carrying parasites capable of sickening humans

(d) damage property

32. An artesian aquifer has a surface area of 6.0 km². Every day, 1,000 m³ are drawn from the aquifer. During a 3-day period, the aquifer experiences no recharge, and the gauge pressure at the top of the aquifer changes from 0.10 atm to 0.09 atm. The aquifer's storage coefficient is most nearly

   (a) 0.00000049
   (b) 0.000015
   (c) 0.0016
   (d) 0.0048

33. Water exits the permeameter shown in the figure at atmospheric pressure. The hydraulic conductivity of the soil in the permeameter is most nearly

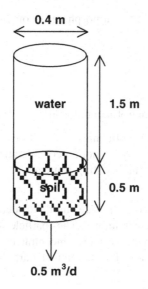

   (a) 1.1×10^{-5} m/s
   (b) 1.5×10^{-5} m/s
   (c) 3.1×10^{-5} m/s
   (d) 3.8×10^{-5} m/s

34. A circular berm with 45° sides is built as an external liner around a 40 m³ tank containing hazardous waste. The berm is to be left uncovered. A 25-year, 24-h precipitation event for the location of the tank is 16 cm. The berm will be 1 m in height. The diameter of the berm halfway between its inner rim and its peak will most nearly be

   (a) 7.3 m
   (b) 7.8 m
   (c) 8.0 m
   (d) 53 m

35. A power surge of sufficient size would disrupt a process that generates revenue of $1,000/h. The disruption would last until the electronic control panel can be

replaced, which takes 2.5 h. The electronic control panels are $500 each. The expected occurrence of power surges sufficient to damage the panels is 0.5/year. A surge protector would protect the control panel in most cases; it is estimated that the control panel would require replacing due to a power surge only once every 10 years if the surge protector were in place. The surge protector can be leased from the power company for $1,100/year. The return on security investment is most nearly

(a) −$800/year
(b) $100/year
(c) $1,200/year
(d) $3,000/year

36. Nearly all of the water in the atmosphere is contained in the

(a) stratosphere
(b) troposphere
(c) mesosphere
(d) thermosphere

37. Maximum achievable control technologies

(a) apply to hazardous air pollutants
(b) are determined by taking economic considerations into account
(c) are different for new and existing sources
(d) all of the above

38. Two optical particle counters are used to determine the particle size distribution of the same sample. The particle distribution from optical particle counter A has more particles concentrated in the $PM_{2.5}$ range than the particle distribution obtained from optical particle counter B. A plausible explanation is that

(a) optical particle counter B has more bends in the sample tubing than optical particle counter A
(b) optical particle counter A has more bends in the sample tubing than optical particle counter B
(c) a larger portion of the sample was analyzed by optical particle counter B than optical particle counter A
(d) a larger portion of the sample was analyzed by optical particle counter A than optical particle counter B

39. A tank contains, at most, 10,000 kg of propylene oxide, which has a density of 0.85 g/cm^3. A diked area will be built that can contain the entire contents of the tank in order to reduce the potential off-site consequences in the event of tank failure. The area to be diked is 100 m^2. The height of the dike must most nearly be

(a) 10 cm
(b) 12 cm
(c) 20 cm
(d) 120 cm

40. The toxicity characteristic leaching procedure (TCLP) is intended to

(a) identify chemicals that are mutagenic or carcinogenic or that are endocrine disrupters
(b) model the process whereby toxic compounds are extracted from their original matrix into landfill leachate
(c) quantify the potential of a toxic material to leach through landfill liner material
(d) identify the degree to which a toxic compound passes through human skin

41. A dose of a toxic compound is fed to a population. The percentage of the population that experiences the average degree of response to the dose, plus or minus the standard deviation of the average degree of response to the dose, is likely to include most nearly

(a) 33 % of the population
(b) 68 % of the population
(c) 95 % of the population
(d) 100 % of the population

42. The level of analysis required by the National Environmental Policy Act that involves the greatest degree of effort happens when

(a) an environmental assessment is needed
(b) an environmental impact statement is needed
(c) an environmental site assessment is needed
(d) categorical exclusion is granted

43. The temperature at ground level is 25 °C. Conditions are typical. One would expect the temperature at a height 1 km above ground level to be most nearly

(a) 12 °C
(b) 19 °C
(c) 25 °C
(d) 32 °C

44. A turbine at a chemical processing plant emits noise that at a distance of 0.5 m measures 125 dB. Workers without ear protection may be exposed to this sound over an 8 h period if they are at least

(a) 1.6 m from the turbine
(b) 2.9 m from the turbine
(c) 8.9 m from the turbine
(d) 28 m from the turbine

45. The heats of formation, ΔH_f, of solid naphthalene ($C_{10}H_8$), carbon dioxide, and liquid water are 36.0, -94.1, and -68.3 kcal/mol, respectively. The higher heating value (HHV) of naphthalene, estimated using the equation for combustion, is most nearly

   (a) $-1,250$ kcal/mol
   (b) $-1,180$ kcal/mol
   (c) -198 kcal/mol
   (d) $1,180$ kcal/mol

46. Concentrations of a compound for one 8 h time period are given in the table. The 8 h time-weighted average concentration of this compound is most nearly

   (a) 27 ppm
   (b) 30 ppm
   (c) 37 ppm
   (d) 45 ppm

47. A solution has a pH of 4. The mass in grams of solid sodium hydroxide (NaOH) needed to neutralize 20,000 L of the solution is most nearly

Time, h	Concentration, ppm
0–1	85
1–2	5
2–3	62
3–3.5	14
3.5–4	2
4–6	78
6–7	6
7–7.5	35
7.5–8	46

   (a) 2 g
   (b) 80 g
   (c) 3,200 g
   (d) 3,200,000 g

48. Henry's Law constant for dimethyl sulfide is 7.1 L-atm/mol. At atmospheric pressure, the concentration of dimethyl sulfide in air that is in equilibrium with a concentration of 2 ppm dimethyl sulfide in water is most nearly

   (a) 1.4×10^{-5} ppm
   (b) 2.3×10^{-4} ppm
   (c) 14 ppm
   (d) 230 ppm

For problems 49 and 50, a ground water aquifer is at a pressure of 1.5 atmospheres and has a temperature of 15 °C. The landfill above it has landfill gas which is saturated with water and consists of 50 % CO_2, 25 % CH_4, and 25 % N_2. Henry's law constants for CO_2 are 1.05×10^3 and for CH_4 is 3.00×10^4 atm/mole fraction.

49. The concentration of CH_4 in the groundwater is most nearly equal to

(a) 0.0001 g/l
(b) 0.011 g/l
(c) 0.11 g/l
(d) 1.00 g/l

50. The concentration of CO_2 in the water is most nearly equal to

(a) 10×10^{-3} g/l
(b) 8.6×10^{-3} g/l
(c) 10.0×10^{-3} g/l
(d) 10.6×10^{-3} g/l

51. A city of 155,000 persons generates waste @ 2.5 kg/day-person. The city's proposed landfill has an area of 55 acres, and the compacted material is expected to have a density of 400 kg/m³. The population is expected to grow @ 1.1 % per year. The landfill can be filled up to 10 m with 20 % of the volume used for cover materials. The expected life of the landfill is most nearly

(a) Landfill cannot be used even for 1 year
(b) 10 years
(c) 20 years
(d) 30 years

52. It is planned to build a sanitary landfill to serve a community of 400,000 for 25 years. The per capita rate of solid waste in the community is 12 kg/person-day. The depth of the landfill is expected to be 15 m, and it is expected that the layers of covers will occupy 15 % of the volume of the landfill. The waste brought to the landfill will be compacted to a density of 2,000 kg/m³. The m² of the land needed will be most nearly

(a) 1.00×10^6
(b) 1.50×10^6
(c) 1.75×10^6
(d) 2.00×10^6

53. A community of 100,000 produces solid waste at 1,500 lb/capita/year. The wastes are compacted to 700 lb/yd³. The depth of the landfill is 10 ft, and 20 % of the volume is taken up by covers. The acres of land needed for 10 years are most nearly

(a) 100
(b) 170
(c) 300
(d) 500

54. The maximum depth of leachate in a landfill is 6.00 ft. The specific quantity of leachate is 1.02. The coefficient of permeability of the clay liner is 0.048 gpd/ft^2 per unit hydraulic gradient. The thickness of the clay liner that will limit the seepage of leachate to 0.11 gpd/ft^2 is most nearly

    (a) 1.02 ft
    (b) 2.30 ft
    (c) 4.60 ft
    (d) 6.00 ft

55. The kg of NH_3 released during the complete aerobic decomposition of a solid waste with formula, $C_5H_{12}O_2N$, per kg of solid waste is most nearly

    (a) 0.01
    (b) 0.15
    (c) 1.00
    (d) 3.00

56. During the anaerobic transformation of a solid waste given by the formula

$$C_{60}H_{90}O_{30}N + xH_2O \rightarrow 35CH_4 + 25O_2 + yNH_3$$

The volume of methane generated per kg of solid waste (at 25 °C and 1 atm pressure) is most nearly

    (a) 0.655 m^2
    (b) 0.800 m^2
    (c) 1.00 m^2
    (d) 1.500 m^2

57. The approximate energy content of a solid waste whose average composition in $C_{10}H_{12}SO$ is most nearly

    (a) 10,000 BTU/lb
    (b) 13,800 BTU/lb
    (c) 16,000 BTU/lb
    (d) 20,000 BTU/lb

58. The thickness of a clay liner in a landfill is 1.5 m. The surface area is 100,000 m^2. If the hydraulic conductivity of the line is 8.0×10^{-9} m/s and the hydraulic head is 1 m, the mass flow rate of leachate of density 1.12 g/cm^3 is most nearly

    (a) 1×10^{-1} kg/s
    (b) 2×10^{-1} kg/s
    (c) 4×10^{-1} kg/s
    (d) 6×10^{-1} kg/s

59. A well of diameter 0.5 m has been pumping at a rate of 5,000 m^3/day for a long time. An observation well located 50 m from the pumped well has been

drawn down by 1.5 m and another well at 150 m is drawn down by 1.0 m. The unconfined aquifer is 50 m thick. The hydraulic conductivity K is most nearly

(a) 0.5
(b) 1.0
(c) 36.0
(d) 50.0

60. A flow field is defined by the stream function

$$\chi(x,y) = 4x^3 + 3x^2y + 4xy + 3xy^2 + 10y^3$$

The formula for the velocities $u(x, y)$ is

(a) $4x^3 + 3x^2y$
(b) $-(4x^3 + 3x^2 + 4x + 6xy + 30y^2)$
(c) $-(3x^2 + 4x + 6xy + 30y^2)$
(d) $-(6xy + 30y^2)$

61. An aquifer of gravel and sand with a porosity of 21.5 % and a specific yield of 15.8 % has a cross-sectional area of 11.5 m^2 and a depth of 15 m. The percentage of water contained in the aquifer that could be extracted is most nearly

(a) 100
(b) 73
(c) 37
(d) 21

For problems 62 and 63, the velocity potential function for a flow system is

$$\varphi(x,y) = 4x^2 + 6xy + 4y^2$$

62. The velocity $u(x, y)$ is equal to

(a) $8x + 6y$
(b) $-8x - 6y$
(c) $-4x^2 - 6xy - 4y^2$
(d) $-8 - 4x$

63. The velocity $v(x, y)$ is given by

(a) $6x + 8y$
(b) $6xy + 4y^2$
(c) $-6xy - 4y^2$
(d) $-6x - 8y$

64. The design criteria set for a sewer pile are
Minimum velocity = 2 ft/s
Flow rate = 5 ft^3/s

The diameter of this concrete pipe is 20″. Therefore, the design slope of the pipe is most nearly

(a) 0.0008
(b) 0.008
(c) 0.08
(d) 0.8

For problems 65 and 66, Galena particles are being carried forward in water at 20 °C which goes into a gravity settling chamber of depth 10 m. The density of the particles is 7.5 g/cm^3, and the diameter of the particles varies from 0.25 to 0.025 mm. The density of water is 998 kg/m^3, and the viscosity is 1.05×10^{-3} kg/m-s.

65. Assuming that the particles settle individual with no effect from other particles, the smallest settling time in the chamber is most likely

(a) 4,760 s
(b) 4,000 s
(c) 2,100 s
(d) 125 s

66. The highest settling time in the chamber is most likely equal to

(a) 6,841 s
(b) 4,761 s
(c) 1,008 s
(d) 992 s

For problems 67, 68, and 69, a packed tower is used to remove SO_2 from an air-SO_2 mixture. A very dilute solution is used to dissolve the SO_2. It is transported and fed (at 1 atm) to the top of the tower from a storage tank at 10 ft^3/min at 70 °F. The top of the tower is kept 20 ft above the floor, and the head loss in the 2″ pipe amounts to 1.00 ft. The density of the liquid is 62.3 lb_m/ft^3, and the viscosity is $\mu = 0.982$ cP.

67. The average velocity of the liquid in the pipe is most nearly

(a) 1 ft/min
(b) 10 ft/min
(c) 62.3 ft/min
(d) 429 ft/min

68. The frictional losses in the pipe per minute are equal to

(a) 1.00 lb_f ft/min
(b) 10.00 lb_f ft/min
(c) 6.23 lb_f ft/min
(d) 62.3 lb_f ft/min

69. Water at 70 °F is pumped at a constant rate of 10 ft³/min from a reservoir kept on the floor of the laboratory to the open top of an absorption tower used in a pilot plant. The top of the absorption tower is located 20 ft above the floor, and the frictional looses in the 2″ pipe amount to $1.00 \frac{ft-lb_f}{lb_m}$. The maximum power level developed by the pump is 0.020 hp. The height at which the water level must be kept in the reservoir is most nearly

 (a) 1.00 ft
 (b) 5.32 ft
 (c) 8.32 ft
 (d) 10.3 ft

For problems 70, 71, and 72, a pump takes brine from the bottom of a supply tank and delivers it into the bottom of another tank. The brine level in the discharge tank is 200 ft above that in the supply tank. The line between the tanks consists of 700 ft of Schedule 40, 6″ pipe. The flow rate is 810 gal/min. In the line, there are two gate valves, four standard tees, and four ells. The specific gravity of the brine is 1.18; the viscosity of the brine is 1.20 cP; and the overall efficiency of the pump and the motor is 60 %. The entrance and the exit from the tanks to the pipes is sharp.

70. The head loss due to the contraction of flow at the entrance to the piping system from the storage tank 1 is most nearly

 (a) 0.11 ft
 (b) 0.63 ft
 (c) 2.0 ft
 (d) 4.0 ft

71. The head loss due to expansion at the tank #2 is most nearly

 (a) 1.3 ft
 (b) 7.1 ft
 (c) 12.66 ft
 (d) 9.00 ft

72. The power consumption for the pump is most nearly

 (a) 5 hp
 (b) 10 hp
 (c) 90 hp
 (d) 200 hp

For problems 73 and 74, an organic chemical needs to be air stripped from a dilute solution in water, at 25 °C, in a packed bed. The packed bed has a diameter of 3 m and a height of 14 m. The Henry's law constant for the chemical is 4.12×10^{-4} m³ atm/mole. The molecular weight of the organic chemical is 176, and the values of liquid and air flow rates are 30 m³/min and 2,000 m³/min, respectively. The water has a dilute solution of the chemical, and the molecular weight of water is 18, density of water being 10 kg/m³. The incoming liquid has a concentration of 1,440 μg/m³, and the outgoing liquid has a concentration of 2.5 μg/m³.

73. From the experimental data provided above, the value of $K_L a$ is most nearly equal to

    (a) 30 min^{-1}
    (b) 10 min^{-1}
    (c) 2.5 min^{-1}
    (d) 1.0 min^{-1}

74. The same organic chemical is thought best to be stripped in an identical tower but of height 30 m in the field. If all operating conditions are identical, for flow rates, temperatures, and inlet concentration, and assuming an identical value of $K_L a$ (since the packing is similar), the expected outlet concentration of the chemical in the water is most nearly

    (a) 0.25 µg/m^3
    (b) 0.28 µg/m^3
    (c) 0.40 µg/m^3
    (d) 0.58 µg/m^3

For problems 75 and 76, ammonia is to be stripped from a dilute ammonia water solution by countercurrent operation with air in a packed column. The operation is at 25 °C; the incoming liquid ammonia solution contains 1,500 mg/L of ammonia, and the exit solution contains 2 mg/L of ammonia. The value of Henry's law constant is 2.25×10^{-4} m^3-atm/mol, and the flow rates of liquid and gas are $1.05 \times 10^{-3} \frac{m^3}{min\text{-}m^2}$ and 7.00 m^3/min-m^2, respectively. The flow rates are given on the basis of per unit square meter of cross-sectional area of the column.

75. The number of transfer units of the column are

    (a) 7.0
    (b) 6.7
    (c) 2.5
    (d) 2.0

76. Past experimental data on the columns used for similar purposes have shown that values of $K_L a$ do not vary more than 3 % for similar packings, as long as the flow rates do not change more than a factor of 2. It is desired to reduce the water flow rate to $0.75 \times 10^{-3} \frac{m^3}{min\text{-}m^2}$, while all other parameters (except the exit liquid concentration) are kept the same. The most likely result for the exit liquid concentration for this column is

    (a) No change in exit concentration
    (b) New exit concentration is 0.134 mg/L
    (c) New exit concentration is 0.5 mg/L
    (d) New exit concentration is 0.75 mg/L

For problems 77, 78, 79, 80, 81, and 82, the results of a settling test on a 60-cm-tall cylinder are shown on the adjoining graph. Initial concentration of solids that were settled is 5,100 mg/L. This experiment is to be used for determining the dimensions of a continuous-flow thickener whose bottom concentration is designed to be 25,000 mg/L. The thickener operates at a design flow rate of 0.05 m³/s.

77. The final height of the sludge zone in the experiment is equal to

   (a) 0.25 cm
   (b) 5.10 cm
   (c) 12.25 cm
   (d) 15 cm

78. The time required to obtain the underflow concentration in the designed unit is most nearly

   (a) 10 min
   (b) 18 min
   (c) 30 min
   (d) 42 min

79. The area for thickening in the designed unit is most nearly

   (a) 104 m²
   (b) 180 m²
   (c) 240 m²
   (d) 280 m²

80. The subsidence velocity in the experimental diagram is most nearly

   (a) 3.7×10^{-5} m/min
   (b) 4.8×10^{-4} m/min
   (c) 3.9×10^{-3} m/min
   (d) 2.1×10^{-2} m/min

81. The overflow rate in the design tank is most nearly

   (a) 20×10^{-3} m³/s
   (b) 25×10^{-3} m³/s
   (c) 45×10^{-3} m³/s
   (d) 60×10^{-3} m³/s

82. The required area of the designed tank is most nearly

   (a) 32 m²
   (b) 64 m²
   (c) 68 m²
   (d) 104 m²

For problems 83 and 84, a sand filter of porosity of 0.54 and a particle diameter of 0.8 mm is used for filtration of city water. The length of the filter bed is 1.5 m and the filter bed expands to 1.65 m during the backwashing phase. The density of the sand is equal to 2.62 g/cm^3, and the kinematic viscosity of water is 1.03 × 10^{-2} cm^2/s.

83. The upward velocity of the water during the backwashing phase is most nearly

(a) 11.6 cm/s
(b) 1.16 cm/s
(c) 0.56 cm/s
(d) 0.45 cm/s

84. The head loss is most nearly equal to

(a) 0.52 m of water
(b) 0.62 m of water
(c) 1.17 m of water
(d) 1.62 m of water

For problems 85 and 86, treated sewage from a sewage plant is discharged at the rate of 98 m^3/s into a flowing river. The flow rate of the river is 1,800 m^3/s, and the velocity is 0.15 m/s. The BOD$_5$ of the treated sewage is 210 mg/L, with a K— 0.24/day. Assume that the coefficient of purification of the river is 4.0. Assume river is initially saturated with oxygen at 9.2 mg/L.

85. The ultimate BOD of the river at the mixing point is most nearly

(a) 10.8 mg/L
(b) 15.2 mg/L
(c) 16.1 mg/L
(d) 18.3 mg/L

86. The distance along the river where the critical dissolved oxygen deficit will occur is

(a) 2.0 km
(b) 9.2 km
(c) 10.8 km
(d) 18.2 km

For problems 87 and 88, a conventional activated sludge process treats 1,000 m^3/h of sewage with a 5-day BOD of 300 mg/L. The BOD$_5$ of the outgoing liquid is designed to be 11 mg/L. Assume that the F/M ratio is 0.20, and the sludge volume index (SVI) is 92. The value of X_{SS} = 2,500 mg/L.

87. The overall volume of the tank is most nearly equal to

(a) 100 m^3
(b) 1,000 m^3
(c) 1,400 m^3
(d) 1,450 m^3

88. The return sludge ratio is most nearly

  (a) 0.20
  (b) 0.25
  (c) 0.30
  (d) 0.35

89. An aerobic digester is used for digesting sludge. The volume of fresh sludge is equal to 2.0×10^3 m^3/d, and the digested sludge volume is 0.81×10^3 m^3/d. Assuming a digestion period of 41 days, and a storage period of 25 days, the required volume of the digester is most nearly equal to

  (a) 20×10^3 m^3
  (b) 50×10^3 m^3
  (c) 60×10^3 m^3
  (d) 70×10^3 m^3

90. A town has a population of 50,000 persons. The amount of solids in the digested sludge is 54 g/person-day. The percentage of solids is 7.5 %, and the specific gravity of solids is 1.03. The dry solids loading rate is 105 kg/m^3/year. Assume that the linear dimensions of the drying beds are 10×30 m. The depth of sludge is most nearly

  (a) 3 cm
  (b) 5 cm
  (c) 7 cm
  (d) 14 cm

91. A waste liquid has a 2-day BOD of 105 mg/L at 15 °C. The ultimate BOD of the sample is 400 mg/L. Therefore, the value of the reaction rate constant at 20 °C is equal to

  (a) 0.15/day
  (b) 0.19/day
  (c) 0.25/day
  (d) 0.38/day

92. A waste liquid sample has a 3-day BOD of 195 mg/L at 18 °C. The ultimate BOD is equal to 300 mg/L. Therefore, the value of the 5-day BOD at 25 °C is equal to

  (a) 300 mg/L
  (b) 272 mg/L
  (c) 230 mg/L
  (d) 195 mg/L

93. A waste liquid sample has a reaction rate constant $K = 0.18$/day at 15 °C. The value of the 5-day BOD is 150 mg/L. Therefore, the ultimate BOD at 15 °C is equal to

(a) 400 mg/L

(b) 350 mg/L

(c) 252 mg/L

(d) 150 mg/L

94. A waste sample is diluted 10 times (100 mL sample plus 900 mL distilled water). The reaction rate constant is $K = 0.20$/day. The value of the 5-day BOD of the diluted sample is 300 mg/L. Therefore, the ultimate BOD of the original sample is most nearly

(a) 474 mg/L

(b) 1,000 mg/L

(c) 4,740 mg/L

(d) 5,000 mg/L

95. A liquid sample has a 2-day BOD of 200 mg/L and $K = 0.21$ day^{-1}. The 5-day BOD is most nearly

(a) 583 mg/L

(b) 450 mg/L

(c) 400 mg/L

(d) 380 mg/L

96. A waste water has a reaction rate constant 0.24 at 25 °C and 0.18 at 15 °C. The value of the reaction rate constant at 30 °C is most nearly

(a) 0.18 day^{-1}

(b) 0.24 day^{-1}

(c) 0.30 day^{-1}

(d) 1.54 day^{-1}

97. A waste water sample is poured into a crucible. The volume of sample $= 100$ mL. The mass of the crucible $= 44.647$ g. The sample in the crucible is subjected to evaporation at 103 °C. The mass of the sample plus crucible before and after the evaporation is 44.831 and 44.714, respectively. The crucible is further heated to 600 °C, cooled, and then weighed. The mass of the crucible plus contents $= 44.683$ g. The total solids in the sample are most nearly

(a) 67 mg/L

(b) 446 mg/L

(c) 500 mg/L

(d) 670 mg/L

98. The fixed solids in the sample above are most nearly

(a) 36 mg/L

(b) 72 mg/L

(c) 200 mg/L

(d) 360 mg/L

99. A crucible weighs 46.458 g. 100 mL of a wastewater sample is poured into it. The weight of crucible plus sample equals 47.158 g. After evaporation of water at 103 °C, the crucible and contents weigh 46.941 g. The crucible is again heated to 600 °C and cooled. The mass is now 46.888 g. The fixed solids in the sample are most nearly

(a) 4,300 mg/L
(b) 2,300 mg/L
(c) 1,000 mg/L
(d) 430 mg/L

100. The volatile solids in the liquid in problem 99 are most nearly

(a) 10 mg/L
(b) 53 mg/L
(c) 100 mg/L
(d) 530 mg/L

Exam 1 Solutions

1. The concentration of benzene in solution has to be converted from mol/L to mol fraction because Henry's Law constant is given in units of atm.

$$x_i = C\rho_{H_2O}MW_{H_2O}$$

$$= \left(3 \times 10^{-5} \frac{mol}{L}\right)\left(0.001 \frac{L}{g}\right)\left(1\left(16 \frac{g}{mol}\right) + 2\left(1 \frac{g}{mol}\right)\right)$$

$$= 5.4 \times 10^{-7}$$

Apply Henry's Law.

$$p_i = hx_i$$

$$= (240 \text{ atm})(5.4 \times 10^{-7})$$

$$= 1.3 \times 10^{-4} \text{ atm}$$

Convert to ppm (ideal gas behavior).

$$C \text{ (ppm)} = \frac{10^6 p_i}{p_{total}} = \frac{10^6(1.3 \times 10^{-4} \text{ atm})}{1 \text{ atm}} = 130 \text{ ppm}$$

The answer is (c). Option (a) is incorrect because it is arrived at by not converting from partial pressure to ppm. Option (b) is incorrect because it is arrived at by not converting from mol/L to mole fraction. Option (d) is incorrect because it is arrived at by neither converting from partial pressure to ppm nor converting from mol/L to mole fraction.

2. The standard deviation of the population, σ, is 14 ppm, and the number in the sample, n, is 13. The standard deviation of the sample would depend on which 13 samples were re-evaluated but is likely to be near the standard deviation of the population.

The answer is (c). Options (a) and (b) are incorrect because the standard deviation of the sample of 13 air samples is not likely to be that much lower than the standard deviation of the entire population of air samples. Option (d) is incorrect because the standard deviation of a sample of 13 of the air samples is not likely to be twice that of the standard deviation of the entire population of air samples.

3. Find the values of z for a normal distribution that correspond to $x = \mu \pm 5$ ppm, or 83 and 93 ppm. The population mean, μ, is 88 ppm.

$$z_1 = \frac{x_1 - \mu}{\sigma} = \frac{83 - 88 \text{ ppm}}{14 \text{ ppm}} = -0.36$$

$$z_2 = \frac{x_2 - \mu}{\sigma} = \frac{93 - 88 \text{ ppm}}{14 \text{ ppm}} = 0.36$$

From a standard normal distribution table, find the area under the curve from $z = 0$ to $z = 0.36$. It is 0.1406. This is the same as the area under the curve from $z = 0$ to $z = -0.36$, because the curve is symmetrical around $z = 0$. Thus, the total area under the curve from $z = -0.36$ to $z = 0.36$ is 0.2812. This means that there is a 28 % probability that a sample will be within 5 ppm of the population's mean.

The answer is (c). Option (a) is incorrect because it is obtained without doubling the area under one-half of the curve.

4. For an unflanged hood, the equation for calculating the flow rate is

$$Q = v_h(10x^2 + A)$$
$$= \left(1.5 \, \frac{\text{m}}{\text{s}}\right)\left(10(0.4 \text{ m})^2 + 0.1 \text{ m}^2\right)$$
$$= 2.6 \text{ m}^3/\text{s}$$

The answer is (b). Option (a) is incorrect because it uses the formula for estimating the flow rate of a flanged hood. Option (c) is incorrect because it is obtained by neglecting to square the distance to the release point. Option (d) is incorrect because it is obtained by not converting the area of the hood to square meters.

5. For first-order reactions, the rate constant is

$$k = \frac{0.693}{t_{1/2}} = \frac{0.693}{4 \text{ years}}$$
$$= 0.173/\text{years}$$

After 18 months, the amount of DDT remaining is

$$Q = Q_0 e^{-kt}$$

$$= (4 \text{ kg}) \exp\left(-\left(0.173\frac{1}{\text{years}}\right)(1.5 \text{ y})\right)$$

$$= 3.1 \text{ kg}$$

The answer is (d). Option (a) is incorrect because it is obtained by failing to convert from months to year. Option (b) is incorrect because it assumes one-fourth of the original DDT is removed every year. Option (c) is incorrect because it is obtained by assuming that one-fourth of the remaining DDT is removed every year.

6. To calculate the geometric mean diameter, find the log of each diameter and calculate the mean of the logs. They are -1.4, -0.85, -0.52, -0.51, -0.24, -0.23, -0.22, -0.21, 0.041, 0.041, 0.38, and 0.98. The geometric mean is ten to the power of the mean of the logs of the data points. The mean of the logs of the data points is the sum of the logs divided by the number of data points, so the geometric mean is

$$\text{geometric mean} = 10^{\frac{(-1.4)+(-0.85)+(-0.52)+(-0.51)+(-0.24)+(-0.23)+(-0.22)+(-0.21)+0.041+0.041+0.38+0.98}{12}}$$

$$= 0.59$$

The answer is (b). Option (d) is incorrect because it is the mean of the particle diameters, as opposed to the geometric mean of the particle diameters.

7. The geometric standard deviation is ten to the power of the standard deviation of the logs. The standard deviation of the logs is the square root of the sum of the difference between each log value and the mean log value, divided by the number of data points, so the geometric standard deviation is

$$\text{geometric standard deviation} = 10^{\sqrt{\frac{(-1.4-0.59)^2+(-0.85-0.59)^2+(-0.52-0.59)^2+(-0.51-0.59)^2+(-0.24-0.59)^2+(-0.23-0.59)^2}{+(-0.22-0.59)^2+(-0.21-0.59)^2+(-0.041-0.59)^2+(-0.041-0.59)^2+(-0.38-0.59)^2+(-0.98-0.59)^2}{12}}}}$$

$$= 4.0$$

The answer is (d). Option (a) is incorrect because it is the standard deviation of the logs of the particle diameters, instead of ten to the power of the standard deviation of the logs. Option (c) is incorrect because it is the standard deviation of the diameters, as opposed to the geometric standard deviation.

8. The distribution of these data points most nearly resembles a lognormal distribution. Most of the logs of the points cluster around the log of the geometric mean, and the lowest log value is nearly the same distance from the log of the geometric mean as the highest log value. Tabulating the log values into ranges yields the following result.

Range	Number of data points
−1.5 to −2	0
−1 to −1.5	1
−.5 to −1	3
0 to −.5	4
0–.5	3
0.5–1	1
1–1.5	0

Inspection of this tabulated data shows that the frequency increases around the log of the geometric mean and decreases away from the log of the geometric mean. Graphing these data would result in a bell-shaped curve with the log of the geometric mean near the peak of the bell. Also, one can show that approximately two-thirds of the logs of the data points lie within one log of the geometric standard deviation from the log of the geometric mean, and approximately 95 % of the data points lie within 2 logs of the geometric standard deviations from the geometric mean.

The answer is (a).

9. The percentage of particles that are $PM_{0.1}$ is the percent of particles with aerodynamic diameter equal to or less than 0.1 µm. The particles are perfect spheres so the aerodynamic diameter is the actual diameter. One of the 12 particles has a diameter less than 0.1 µm, so the percentage of particles that are $PM_{0.1}$ is

$$\text{Percentage of } PM_{0.1} = \frac{1}{12}(100\ \%) = 8\ \%$$

The answer is (b). Option (c) is incorrect because it is the percentage of particles that are not $PM_{0.1}$.

10. Dry air is approximately 78 % nitrogen.

The answer is (d).

11. The equation for calculating a compound's atmospheric half-life from its reaction rate with hydroxyl radicals is

$$t_{1/2} = \frac{0.693}{k_{OH}[\cdot OH]}$$

$$= \frac{0.693}{\left(1.3 \times 10^{-13}\ \frac{cm^3}{molecules}\right)\left(8.5 \times 10^5\ \frac{molecules}{cm^3}\right)\left(3600\ \frac{s}{h}\right)\left(24\ \frac{h}{d}\right)}$$

$$= 72\ d$$

The answer is (c). Option (b) is incorrect because it is the inverse of the correct answer. Option (d) is incorrect because it neglects the term in the numerator. Option (a) is incorrect because it is the inverse of option (d).

12. The "photo" in photochemical smog indicates that light is required for some of the reactions to take place. Sunny days, all other things being equal, have less atmospheric stability than non-sunny days. The release of smog-forming pollutants could be negatively or positively impacted by sunny weather.

The answer is (b).

13. Beta particles are made up of electrons or positrons, which cannot penetrate aluminum, lead, or concrete but are capable of penetrating skin.

The answer is (c).

14. Compounds separate in the chromatographer in the order of their size, with the smaller compounds eluting first. n-Hexane, with 6 carbon atoms per molecule, is the second largest compound in this group of compounds. It is the compound with the highest concentration in the sample.

The answer is (c).

15. The equation for combustion efficiency is

$$CE = \frac{[CO_2]}{[CO_2] + [CO]} \times 100 \%$$

Solve for CO.

$$[CO_2] + [CO] = \frac{[CO_2]}{CE} \times 100 \%$$

$$[CO] = \frac{[CO_2]}{CE} \times 100 \% - [CO_2]$$

$$= \left(\frac{140,000 \text{ ppm}}{99.95 \%}\right) \times 100 \% - 140,000 \text{ ppm}$$

$$= 70 \text{ ppm}$$

The answer is (d).

16. Use average ingestion rates, inhalation rates, and body mass for a child. The average soil ingestion rate for children aged 1–6 years is 200 mg/day, while the average intake of drinking water is 1 L/day and the average air intake is 0.46 m^3/h. The average body mass of a child is 10 kg. Get the exposure for each compartment and add them.

$$\text{exposure} = \text{concentration} \times \text{intake rate}$$

$$E_{air} = \frac{\left(0.46\frac{m^3}{h}\right)\left(24\frac{h}{day}\right)\left(2\frac{ng}{m^3}\right)\left(\frac{1}{10^6}\frac{mg}{ng}\right)}{10 \text{ kg}} = 2.2 \times 10^{-6} \text{ mg/kg/day}$$

$$E_{soil} = \frac{\left(200\frac{mg}{day}\right)\left(30\frac{ng}{g}\right)\left(\frac{1}{10^9}\frac{g}{ng}\right)}{10 \text{ kg}} = 6.0 \times 10^{-7} \text{ mg/kg/day}$$

$$E_{water} = \frac{\left(1\frac{L}{day}\right)\left(20\frac{ng}{L}\right)\left(\frac{1}{10^6}\frac{mg}{ng}\right)}{10 \text{ kg}} = 2.0 \times 10^{-6} \text{ mg/kg/day}$$

Parts per million in fish is on a mass basis.

$$E_{fish} = \frac{\left(6.4\frac{kg}{year}\right)\left(10^6\frac{mg}{kg}\right)\left(\frac{1}{365}\frac{year}{day}\right)\left(\frac{1}{10^6}\frac{g}{g}\right)}{10 \text{ kg}} = 1.8 \times 10^{-3} \text{ mg/kg/day}$$

$$E_{total} = 2.2 \times 10^{-6}\frac{mg}{kg \cdot d} + 6.0 \times 10^{-7}\frac{mg}{kg \cdot d} + 2.0 \times 10^{-6}\frac{mg}{kg \cdot d} + 1.8 \times 10^{-3}\frac{mg}{kg \cdot d}$$
$$= 1.8 \times 10^{-3} \text{ mg/kg/d}$$

The answer is (c). Note that exposures other than fish intake are negligible.

17. Use average ingestion rates, inhalation rates, and body mass for an adult. The average soil ingestion rate for adults is 100 mg/d, while the average intake of drinking water is 2 L/d and the average air intake is 0.83 m^3/h. The average body mass of an adult is 70 kg. Get the exposure for each compartment and add them.

$$\text{exposure} = \text{concentration} \times \text{intake rate}$$

$$E_{air} = \frac{\left(0.83\frac{m^3}{h}\right)\left(24\frac{h}{day}\right)\left(2\frac{ng}{m^3}\right)\left(\frac{1}{10^6}\frac{mg}{ng}\right)}{70 \text{ kg}} = 5.7 \times 10^{-7} \text{ mg/kg/day}$$

$$E_{soil} = \frac{\left(100\frac{mg}{day}\right)\left(30\frac{ng}{g}\right)\left(\frac{1}{10^9}\frac{g}{ng}\right)}{70 \text{ kg}} = 4.3 \times 10^{-8} \text{ mg/kg/day}$$

$$E_{water} = \frac{\left(2\frac{L}{day}\right)\left(20\frac{ng}{L}\right)\left(\frac{1}{10^6}\frac{mg}{ng}\right)}{70 \text{ kg}} = 5.7 \times 10^{-7} \text{ mg/kg/day}$$

Parts per million in fish is on a mass basis.

$$E_{fish} = \frac{\left(45.8\frac{kg}{year}\right)\left(10^6\frac{mg}{kg}\right)\left(\frac{1}{365}\frac{year}{d}\right)\left(\frac{1}{10^6}\frac{g}{g}\right)}{70 \text{ kg}} = 1.8 \times 10^{-3} \text{ mg/kg/day}$$

$$E_{total} = 5.7 \times 10^{-7} \frac{mg}{kg \cdot d} + 4.3 \times 10^{-8} \frac{mg}{kg \cdot d} + 5.7 \times 10^{-7} \frac{mg}{kg \cdot d} + 1.8 \times 10^{-3} \frac{mg}{kg \cdot d}$$

$$= 1.8 \times 10^{-3} \; mg/kg/d$$

The answer is (c). Again, the exposures other than fish are negligible. The exposure per body mass is the same as for a child even though much more fish is consumed by adults.

18. Lowering the definition of leak rate increases the annualized cost of the leak detection and repair program by

$$\text{change in annualized cost of program} = \$350{,}000 - \$204{,}000$$
$$= \$146{,}000$$

The value of the product not lost to fugitive emissions as a result of the program is

$$\text{value of product not lost} = (\text{product value})(\text{amount of emission reduction})$$
$$= \left(2.10 \frac{\$}{L}\right)\left(70 \frac{Mg}{year}\right)\left(\frac{1}{0.9}\frac{L}{kg}\right)\left(1{,}000 \frac{kg}{Mg}\right)$$
$$= \$163{,}000/year$$

Therefore, the net annualized cost of lowering the leak definition is

$$146{,}000 \frac{\$}{year} - 163{,}000 \frac{\$}{year} = -\$17{,}000/year$$

The answer is (d). Option (a) is incorrect because it was obtained without correcting for the savings in product that was previously lost to the environment. Option (c) is incorrect because it was obtained without recognizing that there is a cost savings rather than a cost.

19. Destruction and removal efficiency is given by

$$DRE = \frac{W_{in} - W_{out}}{W_{in}} \times 100\;\%$$
$$= \left(\frac{20\frac{kg}{d} - 0.0045\frac{kg}{d}}{20\frac{kg}{d}}\right) \times 100\;\%$$
$$= 99.98\;\%$$

The answer is (d). Option (a) is incorrect because it is arrived at by failing to multiply by 100.

20. Integrated mosquito management includes removal of standing water that is required for the larval stage of the mosquito life cycle. Where standing water

cannot be removed, mosquito-eating fish or applications of mosquito larvicide might be introduced. Spraying of adult mosquitoes can achieve a temporary reduction in the mosquito population and can be a part of an integrated pest management program to reduce mosquitoes.

The answer is (d).

21. The coefficient of variation is given by

$$
\begin{aligned}
CV &= \frac{100\,s}{\bar{x}} \\
&= \frac{100(60\ \text{ppm})}{240\ \text{ppm}} \\
&= 25\,\%
\end{aligned}
$$

The probability of getting a false positive, α, is 20 %, so the confidence level is

$$
\begin{aligned}
\text{confidence level} &= 1 - \alpha \\
&= 1 - 0.2 \\
&= 80\,\%
\end{aligned}
$$

The probability of getting a false negative, β, is 5 %, so power is

$$
\begin{aligned}
\text{power} &= 1 - \beta \\
&= 1 - 0.05 \\
&= 95\,\%
\end{aligned}
$$

Tables are available that give the number of samples required in a one-sided one-sample t-test to achieve a minimum detectable relative difference at various confidence levels and powers; in this case 41 samples are required.

The answer is (b).

22. It should be assumed that ppm for gases is on a molecular count basis. For an ideal gas, parts per million is equivalent to liters per million liters of air. There are 22.4 L of carbon dioxide in each mole at standard temperature and pressure. Therefore,

$$
350\ \text{ppm}\ CO_2\ (\text{volume}) = \left(\frac{350\ \text{L}\ CO_2}{10^6\ \text{L air}}\right)\left(\frac{\text{mol}\ CO_2}{22.4\ \text{L}\ CO_2}\right)
$$

$$
= 1.6 \times 10^{-5}\ \text{mol/L}
$$

The answer is (b). Option (a) is incorrect because it assumes that the ppm is on a mass basis. Option (c) is incorrect because it fails to include a conversion for 22.4 L/mol.

23. A consequence chart shows that if caustics (such as sodium hydroxide) are mixed with halogenated organics (such as perchloroethylene), heat and flammable gas are generated.

The answer is (b).

24. The risk of getting cancer due to a lifetime spent inhaling a chemical in air is

$$risk_{air} = C_{air} \times air\ unit\ risk$$

$$= \left(0.001 \frac{mg}{m^3}\right) \left(0.00026 \frac{m^3}{\mu g}\right) \left(1000 \frac{\mu g}{mg}\right)$$

$$= 0.00026$$

There are 20,000 exposed persons, so the number of excess cancer cases is

$$20,000\ people(0.00026) = 5\ people$$

The answer is (c). Option (a) is incorrect because air unit risk is cancer risk. Option (b) is incorrect because it is obtained by failing to convert μg to mg.

25. The formula for calculating radioactivity over time is

$$N = N_0 \exp\left(-0.693 \frac{t}{\tau}\right)$$

Solve for t.

$$t = -\frac{\tau \ln\left(\frac{N}{N_0}\right)}{0.693}$$

$$= -\frac{(13\ d) \ln\left(\frac{1.0\ Bq}{10\ Bq}\right)}{0.693}$$

$$= 43\ d$$

The answer is (d). Option (a) is incorrect because it does not take the natural log of the activity ratio. Option (b) is incorrect because it omits the 0.693 term.

26. A waste is considered ignitable if it (1) has a flash point less than 140 °F, unless it is an aqueous solution containing less than 20 % alcohol by volume; (2) is a non-liquid that burns vigorously and persistently and that can ignite through friction or through absorption of moisture or can ignite due to spontaneous chemical reactions; (3) is an ignitable compressed gas; or (4) is an oxidizer.

The answer is (b). Option (a) is incorrect because it is an aqueous solution containing less than 20 % alcohol by volume. Option (d) is incorrect because

to be considered ignitable, a liquid has to be able to ignite through friction or absorption of moisture, or through spontaneous chemical reactions.

27. The gauge pressure at the hole is

$$P_{gauge} = \rho g h$$

The mass flow rate of the contents of the tank is

$$\dot{m} = CA_{hole}\sqrt{2\rho P_{gauge}}$$

Substituting yields

$$\dot{m} = CA_{hole}\sqrt{2\rho P_{gauge}} = CA_{hole}\rho\sqrt{2g\Delta h}$$

$$= (0.55)(80 \text{ cm}^2)\left(0.9\frac{\text{g}}{\text{mL}}\right)\left(1\frac{\text{mL}}{\text{cm}^3}\right)\left(100\frac{\text{cm}}{\text{m}}\right)\left(\frac{1}{1000}\frac{\text{kg}}{\text{g}}\right)\sqrt{2\left(9.8\frac{\text{m}}{\text{s}^2}\right)(4 \text{ m})}$$

$$= 35 \text{ kg/s}$$

The answer is (b). Option (a) is incorrect because it is obtained by neglecting the square root of 2 in the formula. Option (c) is incorrect because it is obtained by putting the gravitational constant outside of the square root. Option (d) is incorrect because it is obtained by failing to take the square root.

28. Use a table that gives ratios of air flow rates to cloth area for baghouses. For silica in shaker/woven baghouses, the ratio of air to cloth is 0.8 $(\text{m}^3/\text{min})/\text{m}^2$, or

$$\frac{\text{air flow rate}}{\text{cloth area}} = 0.8\frac{\frac{\text{m}^3}{\text{min}}}{\text{m}^2}$$

Solve for cloth area.

$$\text{cloth area} = \frac{\text{air flow}}{0.8\frac{\left(\frac{\text{m}^3}{\text{min}}\right)}{\text{m}^2}}$$

$$= \frac{\left(0.3\frac{\text{m}^3}{\text{s}}\right)\left(\frac{60 \text{ s}}{\text{min}}\right)}{0.8\frac{\left(\frac{\text{m}^3}{\text{min}}\right)}{\text{m}^2}}$$

$$= 23 \text{ m}^2$$

The answer is (c). Option (a) is incorrect because it is arrived at by failing to convert seconds to minutes. Option (b) is incorrect because it is arrived at by using the air:cloth ratio for a pulse jet/felt baghouse.

29. The rate of increase is the difference between the concentrations, multiplied by the flow rate.

$$r = \left([H^+]_{out} - [H^+]_{in}\right)Q$$

Hydronium ion concentration is related to pH as follows.

$$[H^+] = \frac{1}{10^{pH}}$$

The hydronium ion concentrations in the scrubber fluid inlet and in the scrubber fluid outlet are

$$[H^+]_{in} = \frac{1}{10^9} = 1 \times 10^{-9} \text{ mol/L}$$

$$[H^+]_{out} = \frac{1}{10^8} = 1 \times 10^{-8} \text{ mol/L}$$

The rate of increase in hydronium ion concentration is

$$r = \left(1 \times 10^{-8} \frac{\text{mol}}{\text{L}} - 1 \times 10^{-9} \frac{\text{mol}}{\text{L}}\right)\left(0.2\frac{\text{L}}{\text{s}}\right) = 1.8 \times 10^{-9} \text{ mol/s}$$

The answer is (a). Option (c) is incorrect because it assumes that hydronium ion concentration is equal to the inverse of pH. Option (d) is incorrect because it assumes hydronium ion concentration is equal to pH.

30. Hydrogen chloride has molecular formula HCl. One mole of hydrogen chloride is removed for every mol of hydronium ion that is transferred to the scrubber fluid. In order to get the mass flow rate of hydrogen chloride removal, find the molecular weight of hydrogen chloride.

$$MW_{HCl} = 1\left(1\frac{\text{g}}{\text{mol}}\right) + 1\left(35\frac{\text{g}}{\text{mol}}\right) = 36 \text{ g/mol}$$

The mass flow rate of removal of hydrogen chloride from the gas stream is

$$r = \left(1.8 \times 10^{-9}\frac{\text{mol}}{\text{s}}\right)\left(36\frac{\text{g}}{\text{mol}}\right) = 6.5 \times 10^{-8} \text{ g/s}$$

The answer is (b). Option (a) is incorrect because it does not convert from a molar flow rate to a mass flow rate.

31. Rats and mice are known as vectors because they can have disease-carrying parasites capable of sickening humans.

The answer is (c).

32. The storativity of a confined aquifer is the change in volume of water per unit area per unit change in head, or

$$S = \frac{\Delta V}{A \Delta h}$$

Convert the change in pressure to head

$$\Delta P = \rho g \Delta h$$

$$\Delta h = \frac{\Delta P}{\rho g}$$

Substitute and solve.

$$S = \frac{\Delta V}{A \left(\frac{\Delta P}{\rho g} \right)} = \frac{\Delta V \rho g}{A \Delta P}$$

$$= \frac{\left(1,000 \, \frac{m^3}{day} \right) (3 \text{ days}) \left(1,000 \, \frac{kg}{m^3} \right) \left(9.8 \, \frac{m}{s^2} \right)}{(6 \text{ km}^2) \left(1,000 \, \frac{m}{km} \right)^2 (0.10 - 0.09 \text{ atm}) \left(1.01325 \times 10^5 \, \frac{Pa}{atm} \right)}$$

$$= 0.0048$$

The answer is (d). Option (a) is incorrect because it is arrived at by neglecting to convert the pressure change to head. Option (c) is incorrect because it is arrived at by neglecting to account for three days of withdrawal.

33. The cross-sectional area of the permeameter is

$$A = \frac{\pi d^2}{4}$$

Darcy's equation is

$$Q = -KA \left(\frac{dH}{dx} \right)$$

Solve for the hydraulic conductivity.

$$K = -\frac{Q}{A \frac{dH}{dx}}$$

$$= -\frac{\left(0.5 \, \frac{m^3}{day} \right) \left(\frac{1}{24} \, \frac{day}{h} \right) \left(\frac{1}{3,600} \, \frac{h}{s} \right)}{\left(\frac{\pi}{4} \right) (0.4 \text{ m})^2 \left(\frac{0 - 1.5 \, m}{0.5 \, m} \right)}$$

$$= 1.5 \times 10^{-5} \text{ m/s}$$

The answer is (b). Option (c) is incorrect because it is arrived at by neglecting to divide by the length of the soil when determining dH/dx. Option (d) is incorrect because it is arrived at by assuming that 0.4 m is the radius of the permeameter, not the diameter.

34. The berm is to be left uncovered so it must be large enough to contain the entire contents of the tank plus the volume of precipitation that can gather within the berm during a 25-year, 24-h precipitation event. Calculate the volume of the berm, where D is the diameter of the berm halfway between its peak and its inside bottom edge.

$$V_{berm} = \pi \left(\frac{D}{2}\right)^2 h_{berm} = 0.25 \text{ m } \pi D^2$$

$$= V_{tank} + V_{rain} = V_{tank} + \pi \left(\frac{D+1 \text{ m}}{2}\right)^2 h_{rain}$$

$$= 40 \text{ m}^3 + \pi \left(\frac{D+1 \text{ m}}{2}\right)^2 (16 \text{ cm}) \left(\frac{\text{m}}{100 \text{ cm}}\right)$$

$$= 40 \text{ m}^3 + (0.04\pi \text{ m})(D^2 + 2 \text{ m } D + 1 \text{ m}^2)$$

Solve for D.

$$0.25 \text{ m } \pi D^2 - (0.04\pi \text{ m})(D^2 + 2 \text{ m } D + 1 \text{ m}^2) = 40 \text{ m}^3$$

$$D = \frac{0.08\pi \text{ m}^2 + \sqrt{(0.08\pi \text{ m}^2)^2 - 4(0.21\pi \text{ m})(-40 \text{ m}^3 - 0.04\pi \text{ m}^3)}}{2(0.21\pi \text{ m})} = 8.0 \text{ m}$$

The answer is (c). Option (a) is incorrect because it is obtained without adjusting for rainfall collected within the berm during a 25-year, 24-h event. Option (b) is incorrect because it does not account for the rain that falls on the outer 0.5 m of the area enclosed by the berm. Option (d) is incorrect because it is equal to the square of the diameter.

35. The return on security investment is the current annualized loss expectancy less the proposed annualized loss expectancy less the annual cost of the control measure.

$ROSI = ALE_0 - ALE_{final} - \text{annual cost of control}$

$$= \left(\frac{0.5}{\text{year}}\right)\left(\left(\frac{\$1,000}{\text{h}}\right)(2.5 \text{ h}) + \$500\right) - \left(\frac{0.1}{\text{year}}\right)\left(\left(\frac{\$1,000}{\text{h}}\right)(2.5 \text{ h}) + \$500\right) - \frac{\$1,100}{\text{year}}$$

$$= \$100/\text{year}$$

The answer is (b). Option (a) is obtained by subtracting the cost of the surge protector from the annual loss expectancy after the surge protector is in place. Option (c) is obtained by subtracting the annual loss expectancy after the surge protector is in place form the annual loss expectancy before the surge protector is in place. Option (d) is the annual loss expectancy without the surge protector in place.

36. Nearly all of the water in the atmosphere is contained in the layer nearest the earth: the troposphere.

   The answer is (b).

37. MACTs apply to hazardous air pollutants, are determined by taking economic considerations into account, and are different for new and existing sources.

   The answer is (d).

38. Bends in the sample tubing would selectively remove large particles and would result in an optical counter having enriched $PM_{2.5}$ results. The portion of the sample that is analyzed by either optical counter should make no difference in the size distribution results.

   The answer is (b).

39. The volume of the containment area must be large enough to hold 10,000 kg of propylene oxide and is given by

$$V = Ah = \frac{m}{\rho}$$

   Solving for h, the height of the dike, gives

$$h = \frac{m}{\rho A}$$

$$= \frac{10{,}000 \text{ kg}}{\left(0.85 \frac{\text{g}}{\text{cm}^3}\right)\left(\frac{1}{1{,}000} \frac{\text{kg}}{\text{g}}\right)\left(100 \frac{\text{cm}}{\text{m}}\right)^2 (100 \text{ m}^2)}$$

$$= 12 \text{ cm}$$

   The answer is (b). Option (a) is incorrect because it is arrived at by failing to take density into account. Option (d) is incorrect because it is arrived at by failing to convert units of density.

40. The TCLP is an extraction procedure that was designed to mimic the potential for landfill leachate to extract certain compounds from their original matrix.

   The answer is (b).

41. The degree of response to a dose of a toxic compound in a population generally follows a normal distribution, with some individuals exhibiting no response, some individuals exhibiting the most severe response, and more individuals exhibiting an average response. For a normal distribution, 68 % of the data points lie within one standard deviation of the mean.

   The answer is (b). Option (c) includes individuals lying within two standard deviations of the mean.

42. A categorical exclusion is granted when an environmental assessment and environmental impact statement are not necessary in order to determine that

no significant impacts are expected. An environmental assessment is conducted in order to determine whether an environmental impact statement is necessary. If an environmental assessment indicates that no significant impacts are expected, then a finding of no significant impact (FONSI) is prepared. Otherwise, an environmental impact statement must be prepared. Environmental site assessments, which evaluate the environmental conditions at the site, are used during the preparation of environmental assessments and environmental impact statements.

The answer is (b).

43. The average lapse rate in the troposphere is $r = -6$ to $-7\ °C/km$. At $h = 1\ km$ above ground level, the temperature is expected to be approximately

$$T = T_0 + hr = 25\ °C + 1\ km\left(\frac{-6.5\ °C}{km}\right) = 19\ °C$$

The answer is (b). Option (a) is incorrect because it uses a lapse rate of $-13\ °C/km$, which is not typical.

44. The allowable exposure for an 8-hr period is 90 dB. The change in sound pressure level with distance from a point source is

$$\Delta SPL = 10\ \log_{10}\left(\frac{r_1}{r_2}\right)^2$$

Solve for r_2.

$$r_2 = \frac{r_1}{\sqrt{10^{\frac{SPL_2 - SPL_1}{10}}}} = \frac{0.5\ m}{\sqrt{10^{\frac{90-125}{10}}}} = 28\ m$$

The answer is (d). Option (a) is the distance that would be allowable for a 15 min period (exposure to 120 dB), and option (c) is the distance that would be allowable for a 2 h period (exposure to 100 dB). Option (b) is incorrect because it is the result obtained by taking the exponent of the natural number rather than 10.

45. The balanced equation for complete combustion of naphthalene is

$$C_{10}H_8 + 12O_2 \rightarrow 10CO_2 + 4H_2O$$

$$HHV = \sum_i \Delta H_{f,i}$$

$$= -36\frac{kcal}{mol} + 10\left(-94.1\frac{kcal}{mol}\right) + 4\left(-68.3\frac{kcal}{mol}\right)$$

$$= -1,250\ kcal/mol$$

The answer is (a). Option (b) is incorrect because it is obtained by forgetting to subtract the heat of formation of reactants in an equation. Option (c) is incorrect because it is obtained without balancing the equation for combustion, so that a 1:1 molar ratio is assumed throughout.

46. The time-weighted average is obtained by taking the concentration for each time period, multiplied by the duration of that concentration, divided by the total time period.

$$TWA = \left(\frac{1}{8\ h}\right)\left(\begin{array}{l}(1\ h)(85\ ppm) + (1\ h)(5\ ppm) + (1\ h)(62\ ppm) \\ +(0.5\ h)(14\ ppm) + (0.5\ h)(2\ ppm) + (2\ h)(78\ ppm) \\ +(1\ h)(6\ ppm) + (0.5\ h)(35\ ppm) + (0.5\ h)(46\ ppm)\end{array}\right)$$

$$= 45\ ppm$$

The answer is (d). Option (c) is incorrect because it is obtained by averaging the 9 values for concentration.

47. First find the concentration of hydrogen ions in the solution. This is equal to

$$[H^+] = \frac{1}{10^{pH}}$$
$$= \frac{1}{10^4}$$
$$= 0.0001\ mol/L$$

The molecular weight of NaOH is

$$MW_{NaOH} = 1\left(23\frac{g}{mol}\right) + 1\left(16\frac{g}{mol}\right) + 1\left(1\frac{g}{mol}\right) = 40\ g/mol$$

Also, NaOH donates one hydroxide ion per molecule, and one hydroxide ion neutralizes on hydronium ion. The mass of NaOH required is

$$m_{NaOH} = V[H^+]MW_{NaOH}$$
$$= (20{,}000\ L)\left(0.0001\frac{mol}{L}\right)\left(40\frac{g\ NaOH}{mol}\right)$$
$$= 80\ g\ NaOH$$

The answer is (b). Option (a) is incorrect because it does not take the molecular weight of NaOH into account. Option (d) is incorrect because it is obtained without converting pH to molar concentration.

48. Henry's Law is

$$py_i = hx_i$$

The units of the Henry's Law constant dictate that the units of concentration in the aqueous phase be in mol/L. The molecular weight of dimethyl sulfide is

$$MW_{(CH_3)_2S} = 2\left(12\frac{g}{mol}\right) + 6\left(1\frac{g}{mol}\right) + 1\left(32\frac{g}{mol}\right) = 62 \text{ g/mol}$$

Concentration units of ppm for aqueous solutions are assumed to be on a mass basis unless otherwise specified. Convert ppm to mol/L.

$$x_i = \left(\frac{2 \text{ g } (CH_3)_2S}{10^6 \text{ g } H_2O}\right)\left(\frac{\left(1\frac{g}{mL}\right)\left(1,000\frac{mL}{L}\right)}{62\frac{g}{mol}}\right) = 3.2 \times 10^{-5} \text{ mol/L}$$

Solve for y_i.

$$y_i = \frac{hx_i}{p}$$

$$= \frac{\left(7.1\frac{L \text{ atm}}{mol}\right)\left(3.2 \times 10^{-5}\frac{mol}{L}\right)}{1 \text{ atm}}$$

$$= 2.3 \times 10^{-4}\frac{atm}{atm} = 2.3 \times 10^{-4} \text{ mol fraction}$$

For air, units of ppm are assumed to be on a molar basis unless otherwise specified. Convert mol fraction to ppm.

$$2.3 \times 10^{-4} \text{ mol fraction}\left(10^6\frac{ppm}{mol \text{ fraction}}\right) = 230 \text{ ppm}$$

The answer is (d). Option (a) is incorrect because it is arrived at by not converting ppm (mass) to molar concentration for the aqueous solution. Option (b) is incorrect because it is arrived at by not converting molar fraction in the air to ppm. Option (c) is incorrect because it is arrived at by not converting ppm (mass) to molar concentration for the aqueous solution or molar fraction in air to ppm.

49. Vapor pressure of water at 15 °C is 1.7051 kPa

$$= 0.017051 \times 760 \text{ mm Hg}$$
$$= 12.95 \text{ mm Hg}$$

The partial pressure of CH_4 is p_{CH4}

$$= \frac{0.25 \times ((1.5 \times 760) - 12.95)}{760} \text{ atm}$$
$$= \frac{0.25 \times 1127}{760} \text{ atm}$$
$$= 0.37 \text{ atm}$$

The mole fraction of CH_4 in groundwater

$$= \frac{p_{CH_4}}{\text{Henry's law constant}}$$

$$= \frac{0.37 \text{ atm}}{3.00 \times 10^4 \text{ atm/mole fraction}} = 0.123 \times 10^{-4} \text{ mole fraction}$$

Let N_{CH_4} be the moles of CH_4 present in 1 L of solution. A liter of solution contains 1,000 g of water

$$= \frac{1,000}{18} \text{ moles } H_2O = 55.56 \text{ g.moles water}$$

$$\therefore 0.123 \times 10^{-4} = \frac{N_{CH_4}}{N_{CH_4} + 55.56}$$

$$\therefore 0.123 \times 10^{-4} N_{CH_4} + 55.56 \times 0.123 \times 10^{-4} = N_{CH_4}$$

$$\therefore N_{CH_4} = \frac{55.56 \times 0.123 \times 10^{-4}}{(1 - 0.123 \times 10^{-4})}$$

$$\approx 6.85 \times 10^{-4} \text{ g.moles}$$

$\therefore$ the concentration of CH_4 is equal to $= (6.85 \times 10^{-4})(16)$
$$= 0.0109 \text{ g/L}$$

$$= (6.85 \times 10^{-4})(16)$$
$$= 0.0109 \text{ g/L}$$

The answer is (b).

50. Vapor pressure of water at 15 °C is $= 12.95 \text{ mmHg}$

$\therefore$ the partial pressure of CO_2 is $p_{CO_2} = \dfrac{0.25 \times ((1.5 \times 760) - 12.95)}{760}$

$$= 0.37 \text{ atm}$$

$\therefore$ mole fraction of CO_2 in groundwater $= \dfrac{p_{CO_2}}{\text{Henry's law constant}}$

$$= \frac{0.37 \text{ atm}}{1.05 \times 10^4 \text{ atm/mol fraction}}$$

$$= 3 \times 10^{-5} \text{ atm/mol fraction}$$

Let N_{CH_4} moles of CO_2 be present in 1 L of solution. The water present in solution $= 55.56$ moles.

$$\therefore 3.52 \times 10^{-5} = \frac{N_{CO_2}}{N_{CO_2} + 55.56}$$

$$\therefore N_{CO_2} \cong 1.95 \times 10^{-4} \text{ moles}$$

$\therefore$ concentration of $CO_2 = 1.95 \times 10^{-4}$ g.moles $\times 44$ g/g.moles

$$= 8.58 \times 10^{-3} \text{ g.moles/L}$$

The answer is (b).

51. Volume of material that can be placed in the landfill

$$= \text{area} \times \text{height}$$
$$= 55 \text{ acres} \times (10 \text{ m} - 0.2 \times 10)$$
$$= 55 \text{ acres} \times 8 \text{ m}$$

$$1 \text{ acre} = 43,560 \text{ ft}^2$$

$$= 43,560 \text{ ft}^2 \times \frac{1 \text{ m}^2}{\left(\frac{100}{30}\right)^2 \text{ ft}^2}$$

$$= 43,560 \times 0.09 \text{ m}^2$$

$$= 3,920 \text{ m}^2$$

$\therefore$ volume that can be placed $= 3,920 \text{ m}^2 \times 8 \text{ m}$

$$= 31363 \text{ m}^3$$

$\therefore$ mass that can be placed $= 31,363 \text{ m}^3 \times \text{density}$

$$= 31,363 \text{ m}^3 \times 400 \text{ kg/m}^3$$

$$= 12,545,200 \text{ kg}$$

A tabular solution can be obtained:

Formula used is Mass generated $=$ population $\times 2.5$ kg/d. person $\times 365$ days/year

$$= \text{population} \times 2.5 \times 365 \text{ kg/year}$$

Year #	Population	Mass of waste, kg/year	Capacity remaining at the end of the year
1	155,000	1.414375×10^8	0

The answer is (a).

52. Mass of the material that will be placed in the landfill in 25 years

$$= 400,000 \text{ persons} \times 12 \text{ kg/person-day} \times 365 \text{ days/year} \times 25 \text{ years}$$
$$= 4.38 \times 10^{10} \text{ kg}$$

Let A_{m^2} be the area of the land needed.

$$\therefore 4.38 \times 10^{10} \text{ kg} = (A)(15)(1 - 0.15)(2,000 \text{ kg})$$

$$\therefore A = \frac{4.38 \times 10^{10}}{15 \times 0.85 \times 2,000} = 1.71 \times 10^6 \text{ m}^2$$

The answer is (c).

53. Volume of waste generated in 10 years

$$= 100{,}000 \text{ persons} \times \frac{1{,}500 \text{ lb}}{\text{person--year}} \times 10 \text{ years} \times \frac{1}{700 \text{ lb/yd}^3}$$
$$= 2{,}142857.1 \text{ yd}^3$$
$$= 2{,}142{,}857.1 \text{ yd}^3 \times \frac{(3)^3 \text{ ft}}{\text{yd}}$$
$$= 5.785 \times 10^7 \text{ ft}^3$$

Let A be areas of land needed.

$$\therefore (A)(10)(0.8) \text{ acres-ft} \times 43{,}560 \text{ ft}^2/\text{acre} = 5{,}785 \times 10^7$$

$$\therefore A = \frac{5.785 \times 10^7}{43560 \times 10 \times 0.8} = 166 \text{ acres}$$

The answer is (b).

54.

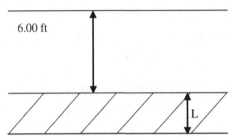

Let L be the thickness of the clay liner.

$$\text{Darcy's equation: } Q = -KA\frac{dh}{dL}$$

$$\text{or : } |Q| = \left| -KA\frac{dh}{dL} \right|$$

$$\text{or : } 0.11 = (0.048)(A = 1)\left(\frac{G+L}{L}\right)$$

Therefore $L = 4.6$ ft.
The answer is (c).

55. The formula equation for the aerobic decomposition of the solid waste is given by

$$C_5H_{12}O_2N + XO_2 \rightarrow 5CO_2 + 4.5H_2O + NH_3$$

$$\text{where, } X = 5 + \frac{4.5}{2} - 2 = 5.25$$

Molecular weight of the solid weight $= (5 \times 12 + 12 \times 1 + 16 \times 2 + 14)$

$$= 60 + 12 + 32 + 14$$

$$= 118$$

Molecular weight of ammonia $= 14 + 3 = 17$

Thus, since 118 kg of solid waste yields 17 kg ammonia, the ammonia yielded by 1 kg of solid waste $= \frac{17}{118} = 0.144$ kg

The answer is (b).

56. Molecular weight of the solids waste $= 60 \times 12 + 90 \times 1 + 30 \times 16 + 14$

$$= 720 + 90 + 480 + 14$$

$$= 1,304$$

Molecular weight of methane $= 12 + 4 = 16$

From the formula, 1,304 kg of solid waste yield $(35 \times 16 = 560)$ kg of methane.

$\therefore$ 1 kg of solid waste yields $\dfrac{560}{1,304} = 0.429$ kg methane.

Applying the ideal gas law

$$PV = nRT$$

P pressure in atmospheres (= 1)
T temperature in K = 298 K
N number of g.mol of methane

$$= \frac{\frac{1,000 \text{ g}}{\text{kg}} \times 0.429 \text{ kg}}{16} = 26.8 \text{ g.mol}$$

$$R = 82.05 \, \frac{(\text{cm}^3)(\text{atmospheres})}{(\text{g.moles})(\text{K})}$$

Therefore, volume of gas $V = \dfrac{26.8 \text{ g.moles} \times 82.05 \frac{\text{cm}^3.\text{atm}}{\text{g.moles K}} \times 298 \text{ K}}{1 \text{ atm}}$

$$= 655,284.12 \text{ cm}^3$$

$$= 655284.12 \text{ cm}^3 \times \frac{1 \text{ m}^3}{(10^2)^3 \text{ cm}^2}$$

$$= 0.655 \text{ m}^3$$

The answer is (a).

57. The following table can be used to obtain the mass percentage of each component

Basis: 1 mole of $C_{10}H_{12}SO$

Component	Moles	Molecular weight	Mass	$\text{Mass \%} = \dfrac{\text{Mass}}{\text{Total mass}} \times 100$
Carbon	10	12	120	66.70
Hydrogen	12	1	12	6.67
Sulfur	1	32	32	17.78
Oxygen	1	16	16	8.89
		Total mass	180	

The Dulong formula can be used for computation

$$\text{Energy content} = 45C + 610\left(H_2 - \frac{1}{8}O_2\right) + 40S + 10N$$

where C, H_2, O_2, S, N represent mass percentage of each component.

$$\text{Therefore, energy content} = (145)(66.70) + 610\left(6.67 - \frac{1}{8} \times 8.89\right) + (40)(17.78)$$

$$= 9,681.6 + 3,391.6 + 711.2$$

$$= 13,784.4 \text{ BTU/lb}$$

The answer is (b).

58. Use Darcy's law to compute velocity of flow

$$\text{Velocity} = \frac{K(\text{hydraulic head})}{\text{thickness}}$$

$$= \frac{8.0 \times 10^{-9} \times 1 \text{ m}}{1.5 \text{ m}} = 5.33 \times 10^{-9} \text{ m/s}$$

The mass flow rate $= (\text{velocity})(\text{Area})(\text{Density})$

$$= 5.33 \times 10^{-9} \times 100,000 \text{ m}^2 \times 1.12 \times 10^3 \text{ kg/m}^3$$

$$= 5.97 \times 10^{-1} \text{ kg/s}$$

The answer is (d).

59.

$$K = \frac{Q\ln\left(\frac{v_1}{v}\right)}{\pi\left(h_1^2 - h^2\right)}$$

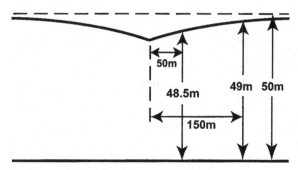

$Q = 5,000$ m³/day
$v_1 = 150$ m
$v = 50$ m
$h_1 = 49$ m
$h = 48.5$ m

$$\therefore K = \frac{5,000 \text{ m}^3/\text{day} \cdot \ln\left(\frac{150}{50}\right)}{\pi\left((49^2) - (48.5)^2\right)} = 35.87 \text{ m/day}$$

The answer is (c).

60.

$$u(x, y) = \frac{-\partial X}{\partial y}$$

$$= \frac{-\partial}{\partial y}\left[4x^3 + 3x^2y + 4xy + 3xy^2 + 10y^3\right]$$
$$= -[0 + 3x^2 + 4x + 6xy + 30y^2]$$
$$= -(3x^2 + 4x + 6xy + 30y^2)$$

The answer is (c).

61. The volume of the aquifer = area × depth = 11.5 m² × 15 m = 172.5 m³.
Since the porosity is 21.5 % and assuming that all pores are completely filled with water

Volume of water in pores = $172.5 \times \frac{21.5}{100} = 37.08$ m³

The specific yield gives the percentage of the aquifer volume that yield extracted water.

Thus, water removed = $172.5 \times \frac{15.8}{100} = 27.25$ m³

∴ percentage of water that is extracted = $\frac{27.25}{37.08} \times 100\% = 73.4$

The answer is b).

62.

$$u(x,y) = \frac{-\partial \varphi}{\partial x}$$

$$= \frac{-\partial \varphi}{\partial x} \left[4x^2 + 6xy + 4y^2 \right]$$

$$= -8x - 6y$$

The answer is (b).

63.

$$V(x,y) = \frac{-\partial \varphi}{\partial y}$$

$$= \frac{-\partial \varphi}{\partial y} \left[4x^2 + 6xy + 4y^2 \right]$$

$$= -6x - 8y$$

The answer is (d).

64. $Q = 5$ ft^3/s; velocity $= 2$ ft/s

Flow rate = (area) (velocity)
Therefore, $Q = Av$
Or

$$A = \frac{Q}{v} = \frac{5 \text{ ft}^3/\text{s}}{2 \text{ ft/s}} = 2.5 \text{ ft}^2$$

$$A_f = \frac{(\pi)(2.5)^2}{4} = 4.909 \text{ ft}^2$$

Therefore $\frac{A}{A_f} = \frac{2.5}{4.909} = 0.5093$
From the graph, $\frac{d}{d_f} = 0.52$
Therefore $d = (0.52)(20'') = 10.4''$
Now velocity $v = \frac{1.486}{n} R^{2/3} S^{1/2}$
Here, $v = 2$ ft/s
$n = 0.013$
Now $\frac{R}{R_f}$ from graph is equal to 98 %

$$R_f = \frac{D}{4} = \frac{2 \text{ ft}}{4} = 0.5 \text{ ft}$$

Therefore $R = (0.98)(0.5 \text{ ft}) = 0.49$ ft
Substituting in the formula, solve for S

$$2 = \left(\frac{1.486}{0.013}\right)(0.49)^{2/3}S^{1/2}$$

Solving for S, the slope $= \left[\frac{(2)(0.013)}{(1.486)(0.49)^{2/3}}\right]^2$

Therefore $S = 0.0079 \approx 0.008$

The answer is (b).

65. The smallest settling time will occur for the fastest settling particles, which are the largest particles of diameter 0.25 mm.

The diameter of the particles is $D_p = 0.25$ mm $= 0.25 \times 10^{-3}$ m
The density of water is $\rho = 998$ kg/m^3
The density of the particles is $\rho p = 7.5$ g/cm$^3 = 7.5 \times 103$ kg/m^3
Viscosity of water is $\mu = 1.05 \times 10^{-3}$ kg/ms

To determine the terminal velocity, one needs to determine the region of the flow (whether Stokes or Intermediate or Newton), which in turn can be found from the K factor.

$$\text{The } K\text{-factor} = Dp \left[\frac{g\rho(\rho_p - \rho)}{\mu^2}\right]^{1/3}$$

$$\text{Therefore, } K = 0.25 \times 10^{-3} \left[\frac{9.8 \times 998 \times (7500 - 998)}{(1.05 \times 10^{-3})^2}\right]^{1/3}$$

Therefore, $K = 9.65$

Since $3.3 < 9.65 < 43.6$, the flow is in the Intermediate low region.

In this region, one can use the terminal velocity formula with $b_1 = 18.5$ and $n = 0.6$.

The terminal velocity is given by

$$u_t = \left[\frac{4gD_p^{n+1}(\rho_p - \rho)}{3b_1\mu^n\rho^{1-n}}\right]^{\frac{1}{2-n}}$$

$$= \left[\frac{4 \times 9.8 \times (0.25 \times 10^{-3})^{0.6+1}}{3 \times 18.5 \times (1.05 \times 10^{-3})^{0.6}(998)^{1-0.6}}\right]^{\frac{1}{2-0.6}}$$

Therefore, $u_t = 0.0794$ m/s.

Thus, the expected settling time in the chamber of depth 10 m is most likely

$$= \frac{10 \text{ m}}{0.0794 \text{ m/x}} = 125.4 \text{ s}$$

The answer is (d).

66. Here again, the K factor needs to be determined to find out the region of flow.

$$D_\rho = 0.025 \times 10^{-3} \text{ m}$$

$$K = Dp\left[\frac{g\rho(\rho_p - \rho)}{\mu^2}\right]^{1/3}$$

Therefore, $K = 0.025 \times 10^{-3}\left[\dfrac{9.8 \times 998 \times (7500 - 998)}{(1.05 \times 10^{-3})^2}\right]^{1/3}$

Therefore, $K = 0.0965$ which puts the flow in the Stokes' law region.
For Stokes' law, $n = 1$, $b_1 = 24$.

Therefore, terminal velocity is $u_t = \left[\dfrac{4gD_\rho^{n+1}(\rho_p - \rho)}{3b_1\,\mu^n\,\rho^{1-n}}\right]^{\frac{1}{2-n}}$

Therefore, $u_t = \left[\dfrac{4 \times 9.8 \times (0.025 \times 10^{-3})^{1+1}(7500 - 998)}{3 \times 24 \times (1.05 \times 10^{-3})^1 (998)^0}\right]^{\frac{1}{2-1}}$

Therefore, $u_t = 0.00210$ m/s
Therefore, settling time $= \dfrac{10\text{ m}}{0.00210\text{ m/s}} = 4761.9$ s
The answer is (a).

67. The average velocity, $\bar{v} = \dfrac{\text{Volumetric flow rate}}{\text{Area of cross section of pipe}}$

Volumetric flow rate $= 10$ ft^3/min
Area of cross section for a 2″ pipe, schedule 40 is equal to 0.02330 ft^2
Therefore, average velocity $= \dfrac{10}{0.0233} = 429.18$
The answer is (d).

68. The frictional losses per lb$_m$ of fluid flow $=$ head loss.

Therefore, friction losses $= 1\dfrac{\text{lb}_f\text{-ft}}{\text{lb}_m}$
Therefore, frictional losses per unit time $=$ (frictional losses per lb$_m$)(mass flow rate lb$_m$/min)

$$= 1\frac{\text{lb}_f\text{-ft}}{\text{lb}_m} \times \frac{10\text{ ft}^3}{\text{min}} \times 62.3\frac{\text{lb}_m}{\text{ft}^3}$$
$$= 62.3 \text{ lb}_f\text{-ft/min}$$

The answer is (d).

69. At 70 °F, the liquid water has the following properties:

$\rho = 62.30$ lb$_m$/ft^3
$\mu = 0.982$ cP

Choose the FPS-gravitational system of units, where

$$1 \text{lb}_f = 32.2 \frac{\text{ft-lb}_m}{\text{s}^2}$$

Use the conservation of mass equation (applied between points a–a and b–b). Here,

$Z_b = 20$ ft

Mass flow rate of water = (vol. flow rate)(density)

$$= \frac{10 \text{ ft}^3}{\text{min}} \times 62.3 \frac{\text{lb}_m}{\text{ft}^3}$$

$$= 623 \frac{\text{lb}_m}{\text{min}}$$

Effective amount of power supplied to the fluid by the pump

$$= 0.20 \text{ hp}$$

$$= 110.0 \frac{\text{ft-lb}_f}{\text{s}}$$

$$= 110 \times 60 \frac{\text{ft-lb}_f}{\text{min}}$$

Therefore, the value of

$$\eta \text{Wp} = \frac{\text{Energy supplied}}{\text{Unit mass of fluid flow}}$$

$$= \frac{\text{Energy supplied/time}}{\text{Mass of fluid flowing/time}}$$

$$= \frac{110 \times 60}{623} = 10.59 \text{ft-lb}_f/\text{lb}_m$$

Since both points a and b are open to the atmosphere, $p_a - p_b = 1$ atmosphere. The total frictional losses in the pipe amount to

$$h_f = 1.00 \text{ ft-lb}_f/\text{lb}_m$$

The velocity of the fluid at point a can be approximately taken as zero (due to the large cross section of the tank). Therefore, $u_a = 0$

At point $b, u_b = \dfrac{\text{Mass flowrate}}{\text{Density} \times \text{Cross-sectional area of pipe}}$

$$= 10 \frac{\text{ft}^3}{\text{min}} \times \frac{1}{\pi \times \left(\frac{2}{12}\right)^2 \times \frac{1}{4} \text{ft}^2}$$

$$= \frac{1}{\pi \times \frac{1}{36} \times \frac{1}{4}} \frac{\text{ft}}{\text{min}}$$

$$u_b = \frac{10 \times 36 \times 4}{\pi \times 60} \frac{\text{ft}}{\text{s}}$$

Therefore, $u_b = 7.636$ ft/s

$$\alpha_a = 1 \ (\text{approx.})$$

One can now susbstitute all of the known terms in the equation of conservation of energy and solve for the unknown term, Z_a

$$\frac{p_a}{\rho} + gZ_a + 0 + 10.59 = \frac{p_b}{\rho} + g(20) + \frac{(1)(7.6)^2}{2 \times 32.2} + 1$$

$p_a = p_b$

Therefore, $Z_a = 20 - 10.59 + \dfrac{(1)(7.6)^2}{2 \times 32.2} + 1$

Therefore, $Z_a = 10.3154$ ft.

Therefore, the fluid in the tank should be located at least 10.3154 ft above the floor level

The answer is (d).

70. The head loss is

$$h_c = \frac{K_c \bar{v}^2}{2g}$$

The fitting coefficient for a sharp entrance to a pipe from a reservoir is

$$K_c = 0.5$$

The inner diameter of schedule 40 6" pipe is 6.065" or .5054 ft. The average velocity in the pipe is

$$\bar{v} = \frac{Q}{A} = \frac{Q}{\frac{\pi}{4}D^2} = \frac{4Q}{\pi D^2}$$

$$h_c = \frac{K_c (\frac{4Q}{\pi D^2})^2}{2g} = \frac{16 K_c Q^2}{2g\pi^2 D^4} = \frac{16(.5)\left(810\,\frac{\text{gal}}{\text{min}}\right)^2 \left(\frac{\text{min}}{60\,\text{s}}\right)^2 \left(\frac{\text{ft}^3}{7.4805\,\text{gal}}\right)^2}{2\pi^2 \left(\frac{32.174\,\text{ft}}{\text{s}^2}\right)(.5054\,\text{ft})^4} = .63\,\text{ft}$$

The answer is (b).

71. The head loss is

$$h_e = \frac{K_e \bar{v}^2}{2g}$$

The fitting coefficient for a sharp exit from a pipe to a reservoir is

$$K_e = 1$$

The inner diameter of schedule 40 6" pipe is 6.065" or .5054 ft. The average velocity in the pipe is

$$\bar{v} = \frac{Q}{A} = \frac{Q}{\frac{\pi}{4}D^2} = \frac{4Q}{\pi D^2}$$

$$h_e = \frac{K_e(\frac{4Q}{\pi D^2})^2}{2g} = \frac{16K_e Q^2}{2g\pi^2 D^4} = \frac{16(1)(810\,\frac{gal}{min})^2(\frac{min}{60\,s})^2(\frac{ft^3}{7.4805\,gal})^2}{2\pi^2(\frac{32.174\,ft}{s^2})(.5054\,ft)^4} = 1.3\,ft$$

The answer is (a).

72. Internal diameter

$$p_a = p_b = 1\,atm$$
$$\rho = 1.18 \times 62.4\,lb_m/ft^3$$
$$\bar{V}_a = \bar{V}_b = 0$$

Frictional: Contraction at A, friction in the pipe (skin friction + form friction due to pipe valves, ells, etc.)

Internal diameter of pipe = 6.065″
$$= 6.065/12\,ft$$
$$= 0.505\,ft$$
Cross-sectional area of pipe = 0.2006 ft²

Area velocity of flow in the pipe is equal to $\dfrac{810}{(7.48)(0.2005)(60)} = 9.00\,ft/s$

$$\frac{K}{D} = \frac{Roughness\ factor}{Diameter} = \frac{0.00015}{0.505} = 0.0003$$

Reynolds number, $N_{Re} = \dfrac{0.505 \times 9.0 \times 1.18 \times 62.4}{1.2 \times 6.72 \times 10^{-4}}$

$$= 4.15 \times 10^5$$

Therefore, from the Moody friction factor chart, $f = 0.0042$
K_f for fittings: 2 gate valves, $2 \times 0.2 = 0.40$
4 elbows, $4 \times 0.9 = 3.60$
4 tees, $4 \times 1.80 = 7.20$
$\Sigma K_f = 11.20$
Contraction loss: $K_c = 0.50$
Expansion loss: $K_e = 1.00$
Total head loss due to friction is, therefore, given by

$$h_f = \left(\frac{0.0042 \times 700}{0.505} + 0.5 + 1.0 + 11.2\right)\left(\frac{9^2}{2 \times 32.2}\right)$$

$$= 23.3\,ft$$

Therefore, using the equation of conservation of energy, total head required is

$$h_t = (200 + 23.3) = 223.3\,ft$$

Mass flow rate of brine $= m = \dfrac{810 \times 8.33 \times 1.18}{60}$

$$= 132.7\,lb_m/s$$

$$\text{Therefore, power consumption} = \frac{223.3 \times 132.7}{550(0.6)} \text{ hp}$$

$$= 89.8 \text{ hp}$$

The answer is (c).

73. The value of $K_L a$ is given by

$$\text{HTU} = \frac{L}{M_w \, K_L a}$$

Therefore, $K_L a = \frac{(\text{HTU})(M_w)}{L}$

where HTU = number of transfer units

M_w = Molar density of water

$$= \frac{1,000 \text{ kg/m}^3}{18 \text{ kg/kg/mole}} = 55.6 \text{ kg/moles/m}^3$$

L = Molar flow rate per square meter of cross section

  = (Volumetric flow rate of liquid)(Molar density)

  = (30)(55.6) kg.moles/min

Therefore, $K_L a = (\text{HTU})(30) = (\text{HTU})(30) \text{ min}^{-1}$

Now, (HTU)(NTU) = height of packed bed

= 14 m

Therefore, $\text{HTU} = \frac{14 \text{ m}}{\text{NTU}}$

NTU can be found from the formula

$$\text{NTU} = \frac{R}{R-1} \ln \left[\frac{\left(\frac{C_{in}}{C_{out}}\right)(R-1) + 1}{R} \right]$$

where C_{in} = incoming concentration

   = 1,440 µg/m^3

C_{out} = outgoing concentration

  = 2.5 µg/m^3

R^1 = Stripping factor, $\left(\frac{H}{R_T}\right)\left(\frac{Q_A}{Q_W}\right)$

Q_A = air flow rate, m^3/min = 2,000 m^3/min

Q_W = water flow rate, m^3/min = 30 m^3/min

R = universal gas constant = $8.206 \times 10^{-5} \frac{\text{m}^3\text{-atm}}{\text{ms/c K}}$

T = temperature K

  = (25 + 273) = 298 K

H = Henry's law constant, m^3-atm/mole

= 4.12×10^{-4} m^3-atm/mole

$$R^1 = \frac{H}{RT}\frac{Q_A}{Q_W} = \frac{4.12 \times 10^{-4}}{8.206 \times 10^{-5} \times 298} \times \frac{2,000}{30}$$

Therefore, $R^1 = 1.12$

$$\text{Therefore, NTU} = \frac{R}{R-1}\ln\left[\frac{\left(\frac{C_{in}}{C_{out}}\right)(R-1)+1}{R}\right]$$

Therefore,

$$= \frac{1.12}{1.12-1}\ln\left[\frac{\left(\frac{1,440}{2.5}\right)(1.12-1)+1}{1.12}\right]$$

$$= 39.66$$

Therefore, HTU $= \dfrac{14\text{ m}}{\text{NTU}} = \dfrac{14}{39.66} = 0.35$

Therefore, $K_L a = (\text{HTU})(30) = 0.35 \times 30 \text{ min}^{-1}$

$$= 10.5 \text{ min}^{-1}$$

The answer is (b).

74. For the experimental data,

$$Z_1 = 14\text{ m} = (\text{HTU})_{exp}(\text{NTU})_{exp}$$

For the field case,

$$Z_2 = 30\text{ m} = (\text{HTU})_{field}(\text{NTU})_{field}$$

Since the two conditions are identical, assume $(\text{HTU})_{field} = (\text{HTU})_{exp}$

Therefore, $(\text{NTU})_{exp} = \dfrac{14\text{ m}}{(\text{HTU})_{field}}$ and $(\text{NTU})_{field} = \dfrac{30\text{ m}}{(\text{HTU})_{field}} = \dfrac{30}{14} \times$ $\dfrac{14}{(\text{HTU})_{field}} = \dfrac{30}{14}(\text{NTU})_{exp}$

Since flow conditions are identical in both cases and Henry's law constant is the same:

Therefore, $R^1_{field} = R^1_{exp} = R^1$

Therefore, $\dfrac{R^1}{R^1-1}\ln\left[\dfrac{\dfrac{C_{in}}{C_{out\,field}}(R^1-1)+1}{R^1}\right]$

$$= \frac{R^1}{R^1-1}\ln\left[\frac{C_{in}}{C_{out\,exp}}(R^1-1)+1\right]\cdot\frac{30}{14}$$

Therefore, $\ln\left[\frac{C_{in}}{C_{out\,field}}(R^1-1)+1\right] = \ln\left[\frac{C_{in}}{C_{out\,exp}}(R^1-1)+1\right]\cdot\frac{30}{14}$

Therefore, taking exponentials of both sides,

$$\frac{C_{in}}{C_{out\,field}}(R^1-1)+1 = e^{\frac{30}{14}} \cdot \left[\frac{C_{in}}{C_{outexp}}(R^1-1)+1\right]$$

But, $R^1 = 1.12$ (computed in problem 87).
$C_{in} = 1,440\ \mu g/m^3$
$C_{out} = 2.5\ \mu g/m^3$
Therefore, $\frac{1,440}{C_{out\,field}}(1.12-1)+1 = (8.52)\left(\frac{1,480}{2.5}\right)(1.12-1)+1)$
Therefore, $\frac{1,440}{C_{out\,field}}(0.12)+1 = 613.78$
Therefore, $C_{out\,field} = \frac{(1,440)(0.12)}{613.78-1} = 0.281\ \mu g/m^3$
The answer is (b).

75. Number of transfer units, NTU $= \frac{R^1}{R^1-1}\ln\left[\frac{\frac{C_{in}}{C_{out}}(R^1-1)+1}{R^1}\right]$

$$R^1 = H^1\frac{Q_A}{Q_W}$$

$Q_A = 7\ m^3/min\text{-}m^2$
$Q_W = 1.05 \times 10^{-3}\ m^3/min\text{-}m^3$

$$H^1 = (H)\left(\frac{1}{RT}\right)$$

$$= \frac{2.25 \times 10^{-4}}{8.206 \times 10^{-5} \times 298}$$

Therefore, $R^1 = \frac{2.25\times10^{-4}}{8.206\times10^{-5}\times298} \times \frac{7}{1.05\times10^{-3}} = 64.4$

$$\frac{C_{in}}{C_{out}} = \frac{1,500}{2} = 750$$

Therefore, NTU $= \frac{64.4}{64.4-1}\ln\left[\frac{750(64.4-1)+1}{64.4}\right]$

$$= 6.72$$

The answer is (b).

76. Since K_La values do not change, the new values of

$$(HTU)_{new} = \frac{(L)_{new}}{M_wK_La} = \frac{(L)_{new}}{(L)_{old}} \cdot \frac{(L)_{old}}{M_wK_La}$$

$$= \frac{(L)_{new}}{(L)_{old}}(HTU)_{old}$$

$$L_{new} = 0.75 \times 10^{-3} \frac{m^3}{min\text{-}m^2}$$

$$L_{old} = 1.05 \times 10^{-3} \frac{m^3}{min\text{-}m^2}$$

Therefore, $(HTU)_{new} = \frac{0.75 \times 10^{-3}}{1.05 \times 10^{-3}} (HTU)_{old} = 0.714(HTU)_{old}$
Thus, since $Z = (HTU)(NTU)$ and Z is kept constant
Therefore, $(HTU)_{new}(NTU)_{new} = (HTU)_{old}(NTU)_{old}$

Therefore, $(NTU)_{new} = \dfrac{(HTU)_{old}(NTU)_{old}}{(HTU)_{new}}$

$$= \frac{(NTU)_{old}}{0.714}$$

New stripping factor $R_{new}^1 = (H^1)\dfrac{(QA)}{(QW)^{new}}$

$$= \left(\frac{H^1 Q_A}{Q_{W\,old}}\right) \cdot \left(\frac{Q_{W\,old}}{Q_{W\,new}}\right)$$

But $\dfrac{Q_{Wold}}{Q_{Anew}} = \dfrac{L_{old}}{L_{new}} = \dfrac{1.05}{0.75} = \dfrac{1}{0.714}$

Therefore, $R_{new}^1 = R_{old}^1 \cdot \dfrac{1}{0.714}$

$R_{old}^1 = 6.44$ (problem #89)

Therefore, $R_{new}^1 = \dfrac{64.4}{0.714} = 90.2$

Therefore, since $(NTU)_{old} = 6.72$ (problem #89)

Therefore, $(NTU)_{new} = \dfrac{6.72}{0.714} = 9.41$

Therefore, since $NTU = \dfrac{R^1}{R^1-1} \ln\left[\dfrac{\frac{C_{in}}{C_{out}}(R^1-1)+1}{R^1}\right]$

And, since $C_{in} = 1,500$ mg/L

Therefore, $9.41 = \dfrac{90.2}{90.2-1} \ln\left[\dfrac{\frac{1,500}{C_{out}}(90.2-1)+1}{90.2}\right]$

$$= 1.011 \ln\left[\frac{\frac{1,500}{C_{out}}(89.2)+1}{90.2}\right]$$

Therefore, $\dfrac{\frac{1,500}{C_{out}}(89.2)+1}{90.2} = \exp\left(\dfrac{9.41}{1.011}\right) = 11,022$

Therefore, $\dfrac{1,500}{C_{out}} = \dfrac{(11,022)(90.2)-1}{89.2} = 11,045$

Therefore, $C_{out} = \dfrac{1,500}{11,045} = 0.134$ mg/L

The answer is (b).

77. The final height of the sludge zone can be found from the formula:

$$H_o C_o = H_u C_u$$

Therefore, $H_u = \dfrac{H_o C_o}{C_u}$

Here, H_o = height of the cylinder = 60 cm
C_o = initial concentration of solids in experiment = 5,100 mg/L
C_u = concentration of bottom solids in designed unit = 25,000 mg/L
Therefore, $H_u = \frac{(60)(5,100)}{(25,000)} = 12.24$ cm
The answer is (c).

78. The time required to obtain the underflow concentration is computed as

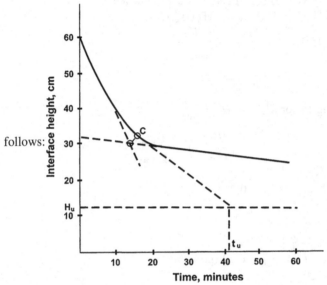

follows:

(1) Locate a horizontal line of height Hu on the graph.
(2) Locate point by bisecting the angle formed by two tangents to the experimental curve.
(3) Draw a tangent at C.
(4) The intersection of the tangent at C and the horizontal Hu gives tu. Thus, tu = 41.5 min.

79.

$$A_t = \frac{(\text{interflow rate})(\text{tu})}{\text{Ho}}$$

$$= \frac{(0.25 \text{ m}^3/\text{s})(41.5 \text{ min})\left(60\frac{s}{\text{min}}\right)}{60 \text{ cm}/\frac{100 \text{ cm}}{\text{m}}}$$

$$= 103.75 \text{ m}^2$$

The answer is (a).

80. The subsidence velocity is in the slope of the left-hand sideline of the graph.

Therefore, $V_S = \frac{60-39}{9.5-0} = \frac{21}{9.5} \frac{cm}{min}$

$$= \frac{21 \; cm}{95 \; min} \times \frac{1 \; min}{60 \; s} \times \frac{1 \; m}{100 \; cm}$$
$$= 3.7 \times 10^{-5} \; m/min$$

The answer is (a).

81.

$$\text{The overflow rate} = \frac{(\text{inflow rate})(\text{height above the sludge zone})}{(\text{initial height})}$$

$$= \frac{0.25 \frac{m^3}{s} \times (60 - 12.5)}{(60)}$$

$$= 0.25 \frac{m}{s} \times \frac{47.5}{60} = 19.8 \times 10^{-3} \; m^3/s$$

The answer is (a).

82. The required area is the larger of two areas:

- The computed area for thickening.
- The computed area for clarification.

The computed area for thickening was computed earlier (problem #93) to be 103.75 m².

The computed area for clarification is given by

$$A_C = \frac{(\text{overflow rate})}{(\text{subsidence velocity})}$$

In problem #94, the subsidence velocity was 3.68×10^{-5} m/min.

In problem #95, the overflow rate was computed to be 19.8×10^{-3} m³/s.

Therefore, the area for clarification $= \frac{19.8 \times 10^{-3} \; m^3/s}{3.68 \times 10^{-5} \; m/min} \times \frac{60 \; s}{1 \; min}$

Therefore, $A_C = 32.2$ m²

The larger of the two areas is $A_t = 103.75$ m², which is the required area in this case.

The answer is (d).

83. The upward velocity, $V = V_S \, \epsilon^{5.0}$

where V_S = settling velocity of sand particles
ϵ = porosity of bed = 0.52 (given)

The settling velocity can be computed with the appropriate formula after obtaining the value for K, the determinant for settling velocity.

$$K = Dp \left[\frac{g\rho(\rho_p - \rho)}{\mu^2} \right]^{1/3}$$

where Dp = diameter of sand particles -0.8 mm = 0.08 cm

g = acceleration due to gravity

   = 980 cm/s^2

ρ = density of water = 1 g/cm^3

ρ_p = density of sand = 2.52 g/cm^3

μ = viscosity of water

$$= \left(\frac{\mu}{\rho}\right)(\rho) = 1.03 \times 10^{-2}\frac{cm^3}{s} \times 1.0 \text{ g/cm}^3$$

$$= 1.03 \times 10^{-2}\frac{g}{cm\text{-}s}$$

Therefore, $K = [0.08]\left[\dfrac{980 \times 1 \times (2.62-1)}{(1.03 \times 10^{-2})^2}\right]^{1/3}$

$$= (0.08)(14.9610^6)^{0.33} = 18.6$$

Here, $3.3 < 18.6 < 43.6$; thus, it falls in the intermediate law region.
For intermediate law region,

$$V_S = \left[\frac{4gD_p^{n+1}(\rho_p - p)}{3b_1\mu^n\rho^{1-n}}\right]$$

where $n = 0.6$, $b_1 = 18.5$ for intermediate law region.

Therefore, $V_s = \left[\dfrac{4 \times 980 \times (0.08)^{1.6}(1.62)}{(3)(18.5)(1.03 \times 10^{-2})^{0.6}(1.00)^{1-0.6}}\right]^{\frac{1}{2-0.6}}$

$$= \left[\frac{4 \times 980 \times 0.0175 \times 1.62}{3 \times 1.85 \times 0.064 \times 1.00}\right]^{\frac{1}{1.4}}$$

$$= 11.6 \text{ cm/s}$$

Therefore, upward velocity of water $= 11.6\frac{cm}{s} \times (0.52)^{5.00}$

$$= 0.44 \text{ cm/s}$$

The answer is (d).

84. The head loss is equal to $L_e (1 - \varepsilon)(\rho_p - 1)$

where L_e = expanded length = 1.5 m

ε = porosity = 0.52

ρ_p = density of sand $- 2.62$ g/cm^3

Therefore, head loss = $(1.5)(1 - 0.52)(2.62 - 1)$ m of water

$$= 1.17 \text{ m of water}$$

The answer is (c).

85. The BOD$_5$ at the mixing point in the river is obtained by material balance on the flow of the sewage and the river.

Total flow after mixing = $1,800 + 98 = 1,898$ m^3/s.

The BOD_5 of the mixture is obtained by material balance

$1,898\ BOD_{5mix} = (1,800)(0) + (98)(210)$

Therefore, $BOD_{5mix} = \frac{(98)(210)}{(1898)} = 10.8$ mg/L

Since the reaction rate constant $K = 0.23$ day^{-1}, the ultimate BOD at the mixing point, L_{mix}, is obtain from

$$BOD_{5mix} = L_{mix}\left(1 - \bar{e}^{K.5}\right)$$

Therefore, $10.8 = L_{mix}\left(1 - \bar{e}^{0.25 \times 5}\right)$

Therefore, $L_{mix} = \frac{10.8}{1 - \bar{e}^{0.23 \times 5}} = \frac{10.8}{1 - \bar{e}^{1.05}} = 18.3$ mg/L

The answer is (d).

86. The travel time from the mixing point to the point of critical deficit is t_c.

The D.O level at the mixing point is $(D.O)_{mix}$.

From $(1,898)(D.O)_{mix} = (1,800)(9.2) = 8.72$ mg/L.

Therefore, $(D.O)mix = \frac{(1800)}{(1898)}(9.2) = 8.72$ mg/L

Thus, initial D.O deficit, D_O, is $= 9.2 - 8.72$

$= 0.48$ mg/L

Formula for t_c is

$$t_c = \frac{1}{K(f-1)}\left[\log\left\{\left\{1 - (f-1)\frac{D_O}{L_{mix}}\right\}f\right\}\right]$$

where $f = 4$.

Therefore, $t_c = \dfrac{1}{0.23(4-1)}\left[\log\left\{1 - (4-1)\dfrac{(0.48)}{(18.3)}\right\}4\right]$

$= \dfrac{1}{0.23 \times 3}\left[4\left\{1 - \dfrac{(3)(0.48)}{(18.3)}\right\}\right]$

$= 0.708$ days

Therefore, the distance along the river $= 0.708$ days

$= 0.708$ days $\times \dfrac{24\text{ h}}{1\text{ day}} \times 3,600\dfrac{\text{s}}{\text{h}} \times 0.15\dfrac{\text{m}}{\text{s}}$

$= 9.17$ Km

The answer is (b).

87. The volumetric inflow rate $Q = 1,000$ m^3/h

$$= 1,000 \times \left(10^2\right)^3 \text{ cm}^3/\text{h} = \frac{1,000 \times \left(10^2\right)^3 \text{ cm}^3/\text{h}}{1,000 \text{ cm}^3/\text{L}}$$

$= 10^6$ L/h $= 0.24$ L/day

If $\frac{F}{M}$ ratio is given by $\dfrac{(Q_{S_I}/VX_{SS})}{1,000}$

where S_I = incoming BOD5
        = 300 mg/L
X_{SS}= 2,500 mg/L
V= volume of tank
$\frac{F}{M}$ = 0.20 (given)
Therefore, $0.20\ V\ = 24 \times 10^6 \times \frac{300}{V} \times \frac{1}{2,500} \times \frac{1}{100}$
Therefore, $V\ = \frac{24}{0.2} \times 10^6 \times \frac{300}{2,500 \times 1,000} = 14,400\ \text{m}^3$
The answer is (d).

88. The return sludge ratio

$$\frac{Q_r}{Q} = \frac{X_{SS}}{\frac{10^6}{SVI} - X_{SS}}$$

Therefore, $R\ = \frac{Q_r}{Q} = \frac{2,500}{\frac{10^6}{92} - 2,500} = 0.298$
The answer is (c).

89. The volume of the digester is V given by

$$V = \left[V_f - \frac{2}{3}\left(V_f - V_d\right)\right]t_1 + V_d t_2$$

where V_f = volume of daily fresh sludge
        = $2 \times 10^3\ \text{m}^3$
V_d = volume of daily dried sludge
      = $0.81 \times 10^3\ \text{m}^3$
t_1 = time of digestion
     = 41 days
t_2 = storage time
     = 25 days
Therefore, $V = \left\{\left[2 - \frac{2}{3}(2 - 0.81)\right](41) + [25][0.81]\right\} \times 10^3$

$$= \left\{\left[\left[2 - \frac{2}{3}(1.91)\right](41)\right] + [(25)(0.81)]\right\} \times 10^3$$
$$= [49.6 + 20.3] \times 10^3$$
$$= 69.9 \times 10^3 = 69,900\ \text{m}^3$$

The answer is (d).

90. The total daily solids = population $\times$ per capita

$$50,000 \times 54\ \text{g/day} = \frac{50,000 \times 54}{1,000}\ \text{kg/day} = 2,700\ \text{kg/day}$$

Volume of digested sludge $= 2,700 \times \frac{100}{7.5} \times \frac{1}{1.03} \times \frac{1}{1,000}\ \text{m}^3/\text{day}$

$$= 34.95\ \text{m}^3/\text{day}$$

Dry solids loading rate = 105 kg/m^3/year

The volume of $\dfrac{\text{solids per year}}{\text{dry solids loading rate}} = \dfrac{2,400 \times 365}{105} = 9,385$ m^2.

The number of drying beds needed = $\dfrac{9,385}{10 \times 30} = 31.2 = 31$

Assume number of cycles per year = 10, with 2 months for repair and down time.

Therefore, depth of sludge = $\dfrac{34.95 \times 365}{32 \times 10 \times 30 \times 10} = 0.132$ m

$= 13.20$ m

The answer is (d).

91. The BOD at any time t is given by

$\text{BOD}_t = L_o(1 - e^{-kt})$

where L_o is the ultimate BOD. From the data, at $t = 2$ days, $\text{BOD}_2 = 105$ mg/L, and $L_o = 400$ mg/L

$\therefore 105 = 400\,(1 - e^{-k \cdot 2})$

$\therefore \frac{105}{400} = 0.263 = 1 - e^{-k \cdot 2}$

$0.737 = e^{-k \cdot 2}$

$\therefore -2\,K = \ln(0.737)$

$\therefore K = 0.15$ day^{-1} at 15 °C

From the literature, $K_{20} = K_{15}\,\theta^{\,20-15}$ where $\theta = 1.047$

$\therefore K_{20} = (0.15)(1.047)^5 = 0.19$ day^{-1}

The answer is (b).

92. The BOD at any time t is given by;

$\text{BOD}_t = L_o\,(1 - e^{-kt})$

where L_o = ultimate BOD

K = reaction rate constant

$\therefore 195 = 300\,(1 - e^{-k \cdot 3})$

$\therefore \frac{195}{300} = 0.263 = 1 - e^{-k \cdot 3}$

$0.35 = e^{-k \cdot 3}$

$\therefore -3K = \ln(0.65)$

$\therefore K = 0.35$ day^{-1} at 18 °C

From the literature, $K_T = K_{18}\,\theta^{\,T-18}$ where $\theta = 1.047$

$\therefore K_{25} = (0.35)(1.047)^{25-18} = 0.48$/day

At 25 °C; $\text{BOD}_5 = L_o\,(1 - e^{-4.8 \times 5})$

$= (300)\,(1 - e^{-2.20})$

$= 272.8$ mg/L

The answer is (b).

93.

$$\text{BOD}_5 = L_o(1 - e^{-k \cdot 5})$$

$$L_o = \frac{\text{BOD}_5}{1 - e^{-5 \cdot k}}$$

Here, $BOD_5 = 150$ mg/L and $K = 0.18$ day^{-1}

$$L_o = \frac{150}{1 - e^{-5 \cdot 0.18}} = 252.7 \text{ mg/L}$$

The answer is (c).

94. For the diluted sample,

$$BOD_5 = L_o(1 - e^{-k \cdot 5}) \text{ or } L_o = \frac{BOD_5}{1 - e^{-5 \cdot k}}$$

For the diluted sample, $K = 0.20$/day and $BOD_5 = 300$ mg/L

$\therefore L_o$ of diluted sample $= \frac{300}{1 - e^{-0.25}} = 473.9$ mg/L

Since the ultimate BOD of distilled water is zero, therefore,

$BOD_{Original} = (BOD_{Diluted})$ (volume diluted sample/volume original sample)

$= (473.9)(1,000/100)$

$= 4,739$ mg/L

$\therefore$ Ultimate BOD of original sample is 4,739 mg/L

The answer is (c).

95. $BOD_t = L_o(1 - e^{-k \cdot t})$

$$\therefore L_o = \frac{BOD_t}{1 - e^{-k \cdot t}}$$

$k = 0.21$ day^{-1}

$t = 2$ days

$\therefore L_o = \frac{300}{1 - e^{-0.25}} = 583.1$ mg/L (Ultimate BOD)

$BOD_5 = L_o (1 - e^{-k \cdot 5})$

$\qquad = (583.1)(1 - e^{-0.21 \cdot 5}) = 379.1$ mg/L

The answer is (d).

96. The value of k is related to temperature by means of Arrhenius law:

$$k = k_o \exp\left(\frac{-\Delta E}{RT}\right)$$

$\therefore \text{Log } k = \log k_e - \left(\frac{\Delta E}{RT}\right)\log_{10}e$, where T is in K

A table of values of k and log k versus $1/T$ is given below:

From the table of values, the equation can be written for the two specific data points:

$$-0.6197 = \log k_o - \left(\frac{(\Delta E)(\log_e 10)}{R}\right)(3.355 \times 10^{-3})$$

Temperature (T)		k, day^{-1}	1/TK (computed)	log k
°C	K			
25	298	0.24	3.355×10^{-3}	-0.6197
15	288	0.18	3.472×10^{-3}	-0.7447

$$-0.7447 = \log k_o - \left(\frac{(\Delta E)(\log_e 10)}{R}\right)(3.472 \times 10^{-3})$$

Subtracting the above two equations,

$$-0.6197 + 0.7447 = \log k_o - \log k_o \left(\frac{(\Delta E)(\log_e 10)}{R}\right)(3.471 - 3.355) \times 10^{-3}$$

$$\therefore 0.1250 = \left(\frac{(\Delta E)(\log_e 10)}{R}\right)(0.117 \times 10^{-3})$$

$$\text{or,} \left(\frac{(\Delta E)(\log_e 10)}{R}\right) = 0.1250/0.117 \times 0^{-3} = 1.068 \times 10^3$$

Substituting this value into the equation for the first data point in the table:
$-0.6197 = \log k_o - 1.068 \times 10^3 \times 13.355 \times 10^{-3}$
$\therefore \log k_o = -0.6197 + 3.5831 = 2.9634$
$\therefore k = 2.9634 \exp\left(\frac{-\Delta E}{RT}\right)$

$$= 2.9634 \exp\left\{\left(\frac{-\Delta E}{RT}\right)(1/T)\right\}$$

Now, $\left(\frac{\Delta E}{R}\right)\log_e 10 = 1.068 \times 10^3$
$\therefore \frac{\Delta E}{R} = 1.068 \times 10^3/\log_e 10 = 463.82$
Thus, $k = k_e \exp\{-463.82/T\}$
At 30 °C i.e. 303 K
$k = 2.9634 \exp\{-463.82/303\} = 2.9634 \exp\{-0.6532\}$
$k = 1.54$ day^{-1}
The answer is (d).

97. Mass of the sample plus the crucible after evaporation at 103 °C, removed of water $= 44.714$ g. The mass of the crucible $= 44.647$ g

$\therefore$ mass of total solids $= 44.714 - 44.647 = 0.067$ g or 67 mg
This is contained in 100 mL of waste water
$\therefore$ Total solids on a per liter basis $= 67$ mg/100 mL $\times$ 1,000 mL/ 1 L $= 670$ mg/L
The answer is (d).

98. The fixed solids constitute the residue left after heating to 600 °C, on a mg/L basis. Thus, the fixed solids in 100 mL of liquid sample

= (44.683 − 44.647) g
= 0.036 g
= 36 mg
Therefore, fixed solids on mg/L basis
= 36 mg/100 mL × 1,000 mL/1L
= 360 mg/L
The answer is (d).

99. The fixed solids are those remaining after heating to 600 °C on mg/L basis. Thus, the fixed solids in 100 mL of liquid = (46.888 − 46.458) g = 0.430 g or 430 mg. Thus, the fixed solids in 1 L of liquid = 430 mg/100 mL × 1,000 mL/1 L = 4,300 mg/L

    The answer is (a).

100. The volatile solids are those that are evaporated between 103 and 600 °C, on mg/L basis. Thus, volatile solids of liquid = (46.941 − 46.888) g = 0.053 g or 53 mg.

    The answer is (d).

Exam 2

For problems 101 through 104, a 100 mL sample of wastewater is heated to 103 °C in a dish that weighs 46.048 g. The mass of the 100 mL of wastewater plus the dish is 46.531 g. After heating to 103 °C, the dish and contents are found to weigh 46.480 g. After heating to 600 °C in an oven and then cooled, the dish and contents weigh 46.424 g. Another empty dish weighing 46.132 g is filled with 100 ml of sample from the same wastewater, but in this case, after being filtered on filter paper. After being heated to 103 °C, the dish and contents weigh 46.494 g. After being heated in an oven to 600 °C, and then allowed to cool slowly, the mass of the dish and the contents is 46.446 g.

101. The total solids in the wastewater are most nearly

  (a) 432 mg/L
  (b) 484 mg/L
  (c) 4,320 mg/L
  (d) 4,848 mg/L

102. The total suspended solids in the wastewater are most nearly

  (a) 70 mg/L
  (b) 700 mg/L
  (c) 3,620 mg/L
  (d) 4,320 mg/L

103. The total volatile solids of the wastewater are most nearly equal to

  (a) 376 mg/L
  (b) 432 mg/L
  (c) 484 mg/L
  (d) 3,760 mg/L

104. The volatile suspended solids in the wastewater are most nearly equal to

  (a) 620 mg/L
  (b) 720 mg/L
  (c) 3,140 mg/L
  (d) 3,760 mg/L

A. Naimpally and K. S. Rosselot, *Environmental Engineering: Review for the Professional Engineering Examination*, DOI: 10.1007/978-0-387-49930-7, © Springer Science+Business Media New York 2013

105. The most broad-based environmental law in the last century is

    (a) Superfund
    (b) Resource Conservation and Recovery Act
    (c) CERCLA
    (d) National Environmental Protection Act (NEPA)

106. The population of a town has been 100,000 in 1990; 112,000 in 1995; and
    125,000 in 2000. The growth has been affected by several factors which are
    expected to remain unchanged till 2050. The Planning Commission has
    decided on setting up a water treatment plan that would be adequate till 2025.
    The per capita water consumption was 40 gallons per day per person in 2000,
    and the water conservation efforts make it likely that the per capita
    consumption will grow at a rate great than 1 % per year. The total water
    consumption per day, which will serve as the design basis for this plant, is
    most nearly

    (a) 7 million gallons per day
    (b) 8 million gallons per day
    (c) 9.3 million gallons per day
    (d) 9.6 million gallons per day

107. A town has a population of 250,000 in year 2000 and is expected to grow at
    the rate of 2 % per year till 2020. The total water consumption in year 2000
    was 10 million gallons per day. The design engineer has come before the city
    council and presented a plan that is based on the expectation that the town
    would consume 16.7 million gallons per day in 2020. A member of the
    council needed to know the percent growth per year in per capita
    consumption of water that was assumed in the calculations of the
    engineer. The engineer did not have the number at hand, and his assistant
    made a quick computation. This computation would yield an answer equal to

    (a) 0.3 %
    (b) 0.6 %
    (c) 1 %
    (d) 1.5 %

108. A round-edged venturimeter with throat diameter $1''$ is placed in a $2''$ pipe.
    The pipe is a horizontal pipe, and the pressure drops 3 mm of mercury. The
    volumetric flow rate is most nearly

    (a) 1.66 m^3/h
    (b) 3.00 m^3/h
    (c) 5.00 m^3/h
    (d) 10.00 m^3/h

109. The stagnation pressure in a pitot tube is equal to 1.1 m of water at a depth of 1 m in river. The velocity of the river at that depth is most nearly

   (a) 1.10 m/s
   (b) 1.41 m/s
   (c) 2.81 m/s
   (d) 4.01 m/s

110. A large water storage tank, 5 m in diameter, has a small hole, 0.5 cm wide, at a distance of 10 m from the top of the tank. The expected amount of water flowing out is

   (a) 10 m³/min
   (b) 17 m³/min
   (c) 162 m³/min
   (d) 1,620 m³/min

For problems 111 through 113, a municipal solid waste sample has the following analysis:

Component	Percentage of mass	Percentage of moisture	Density after compaction lb/ft³	Energy content BTU/lb
Plastics	22	0.5	6.2	14,200
Paper	42	6.8	22.1	9,100
Food waste	18	72.5	81.0	1,800
Cardboard	14	4.1	14.6	7,200
Other	4	14.1	25.2	7,600

111. The moisture content of the solid waste is most nearly (%)

   (a) 22
   (b) 25
   (c) 38
   (d) 81

112. The space required for the landfill for 1 lb of waste is most nearly

   (a) 0.01 ft³
   (b) 0.05 ft³
   (c) 0.07 ft³
   (d) 0.10 ft³

113. The energy content of the solid waste is most nearly

   (a) 14×10^4 BTU/100 lb
   (b) 76×10^4 BTU/100 lb
   (c) 85×10^4 BTU/100 lb
   (d) 10^6 BTU/100 lb

114. A landfill collects methane gas generated in the landfill to generate electricity. However, there is concern that too much methane would be given out to the atmosphere. Assume a porosity of the cover material being 0.215, the landfill cover thickness being 1 m, and the diffusivity of methane as 0.2 cm^2/s. Assume the concentration of methane just below the cover to be 0.001 g/cm^3. The moles of methane given out to the atmosphere per square meter of the landfill cover is

   (a) 2.3×10^{-7} g/s
   (b) 2.3×10^{-3} g/s
   (c) 2.0×10^{-2} g/s
   (d) 0.20 g/s

115. A landfill with a large surface area generates methane and carbon dioxide in the mass ratio of 2:1. The landfill cover has a thickness of 1 m and the porosity of the cover is 0.18. The diffusivities of methane and carbon dioxide are 0.2 cm^2/s and 0.13 cm^2/s, respectively. The ratio of the masses of methane and carbon dioxide given out the atmosphere through the cover is most nearly

   (a) 2:1.0
   (b) 2:1.3
   (c) 4:1.3
   (d) 4:3.0

116. The height of a lift at the bottom of a landfill is initially equal to 5 m. The initial specific weight of the material in the lift is 1,100 lb/yd^3. The relationship between specific weight and pressure is given by

$$SWp = 1,100 + \frac{p}{0.014 + 0.0015\,p}$$

where p is pressure in psi.

The bottom of the landfill is subjected to a pressure of 200 psi. The settlement of the bottom lift will lead to an additional capacity of height equal to (most nearly)

   (a) 1.00 m
   (b) 1.80 m
   (c) 3.00 m
   (d) 5.00 m

117. An operator needs to estimate the kg of methane produced from 1 kg of solid waste undergoing anaerobic decomposition in a landfill. The average formula of the solid waste is $C_{50}H_{70}O_{30}N$. The kg of methane produced per kg of solid waste is most nearly

   (a) 0.015 kg
   (b) 0.255 kg
   (c) 0.355 kg
   (d) 0.484 kg

118. A solid waste is defined under RCRA as

(a) A solid with a melting point of at least 500 °C.

(b) A solid with a melting point of at least 1,000 °C.

(c) A solid or elasto-viscous liquid, i.e., a liquid that has some characteristics of solids.

(d) A discarded liquid, solid, or gaseous material.

119. For problems 119 and 120, a solid waste is collected from a town at 500,000 kg/week. It contains moisture of 45 %. Fifty percent of the solid waste can be converted into compost when mixed with sludge. The sludge available is 200,000 kg/week, which is thickened to 27 % solids and has a C:N ratio of 100:1. The solid waste has a C:N ratio of 15:1. The C:N ratio of the compost is most nearly

(a) 15:1

(b) 25:1

(c) 50:1

(d) 100:1

120. The mass of initial dry mixture which can be converted into compost is most nearly

(a) 100,000 kg/week

(b) 190,000 kg/week

(c) 200,000 kg/week

(d) 500,000 kg/week

121. A water sample contains the following:

Ca^{2+}: 80 mg/L

Mg^{2+}: 106 mg/L

Na^+: 49 mg/L

The total hardness in meq/L is most nearly

(a) 4.0

(b) 13.1

(c) 17.1

(d) 30.2

For problems 122 and 123, the analysis of a water sample is as follows:

Component	Concentration of mg/L
Ca^{2+}	110
Mg^{2+}	100
Na^+	3
Alkalinity	80 as $CaCO_3$
pH	7.2

122. The total hardness of the water sample is most nearly (as mg/L of $CaCO^3$)

    (a) 274
    (b) 411
    (c) 685
    (d) 1,100

123. The non-carbonate hardness of the water (as mg/L of $CaCO_3$) is most nearly

    (a) 80
    (b) 605
    (c) 655
    (d) 685

For problems 124 and 125, in a reverse osmosis unit, the values on the side of the feed stream are
Temperature: 25 °C
Pressure: 50 atmospheres
Concentration of NaCl: 2.2 % $\left(\frac{m}{m}\right)$.
The values on the permeate side are
Temperature: 25 °C
Pressure: 2 atmospheres
Concentration of NaCl: 0.1 % $\left(\frac{m}{m}\right)$.
The values of the $\Delta\pi$, the difference in osmotic pressure, is equal to 8 atmosphere. Values for the permeance for water and NaCl are

water: $W_p = \left(\frac{P_{MH2O}}{l_M}\right) = 5.8 \times 10^{-7}$ g.moles/cm^2 -s-atm

NaCl: $J_s = \left(\frac{P_{MNaCl}}{l_M}\right) = 14.8 \times 10^{-6}$ cm/s

124. The flux of water is most nearly equal to

    (a) 2.3×10^{-5} g/cm^2-s
    (b) 3.5×10^{-5} g/cm^2-s
    (c) 4.1×10^{-4} g/cm^2-s
    (d) 8.0×10^{-4} g/cm^2-s

125. The flux of salt through the membrane is most nearly

    (a) 2.85×10^{-7} g/cm^2-s
    (b) 5.2×10^{-6} g/cm^2-s
    (c) 4.8×10^{-5} g/cm^2-s
    (d) 3.5×10^{-4} g/cm^2-s

126. A flow field has the following velocities:

$$u(x,y) = 4x + 6y$$
$$v(x,y) = -4y$$

with boundary condition of

$$\chi(o,o) = 0$$

The stream function for this flow field given by $\chi(x_1, y)$ which is

(a) $-4xy + 3y^2$
(b) $-4xy$
(c) $3y^2$
(d) 0

127. A confined aquifer 12 m thick has several monitoring wells placed in it. The difference in the level of the piezometric head between two walls is 1.8 m. The two wells have a distance of 1,200 m between them and the value of $K = 22$ m/d. The flow rate in the aquifer is most nearly

(a) 0.1 m^3/day-m width
(b) 0.4 m^3/day-m width
(c) 0.8 m^3/day-m width
(d) 1.0 m^3/day-m width

128. The pump-and-treat process for cleanup is used for

(a) Mixture of groundwater and surrounding soil.
(b) Groundwater which is pumped up, treated, and then returned.
(c) Soil which is cleaned, treated, and water pumped out for recovery of expensive chemicals.
(d) Soil which is removed form the site, cleaned, treated, and chemicals recovered.

129. An artesian well from a confined aquifer pumps out 1,000 m^3/d of drinking water. Two monitoring wells are placed at distances of 50 m and 100 m, respectively, from this well. The aquifer thickness is 50 m and the height of the piezometric surface is 100 m. The dip on the piezometric surface at a distance of 500 m is 1 m and that at 1,000 m is 0.5 m. The value of conductivity K is most nearly

(a) 1.0 m/d
(b) 2.0 m/d
(c) 4.4 m/d
(d) 8.8 m/d

130. An unconfined aquifer has a thickness of 50 m and a drinking water well of capacity $Q = 1,000$ m^3/day. There are two observation wells at distances of 25 and 50 m, respectively. The dip in the level of the water table at these two wells is 0.75 and 0.25 m, respectively. The value of K is most nearly

(a) 4.5 m^2/day
(b) 6.5 m/day
(c) 8.5 m/day
(d) 10.0 m/day

131. A plume of width 80 m is contaminating a confined aquifer of thickness 40 m and $K = 1.5 \times 10^{-2}$ m/min. The hydraulic gradient is 0.0015 and the pumping rate is 0.01 m^3/min. The location of a single extraction well so that it can remove the plume is (on the x-axis) away

(a) 10 m
(b) 20 m
(c) 25 m
(d) 29 m

132. A groundwater well draws $Q = 1,000$ m^3/day from the well. The groundwater aquifer has a thickness of 100 m, a porosity of 30 % and a specific yield of 20 %. The cubic meters of the aquifer "used up" in one day is most nearly

(a) 1,000 m^3
(b) 2,000 m^3
(c) 3,000 m^3
(d) 5,000 m^3

133. It is desired to wash soil containing a petroleum fraction with an average composition of $C_{10}H_{23}$. It is found that 1 kg of the dirty soil (soil plus hydrocarbon) contains 84 % by mass of soil. It is desired to wash the solid with water so that 98 % of the hydrocarbon is removed. If the resultant water/hydrocarbon mixture is made to contain a maximum of 0.01 kg hydrocarbon per kg of dirty water, the minimum mass of water needed to wash 1 kg of dirty soil is most nearly

(a) 0.16 kg
(b) 0.99 kg
(c) 1.50 kg
(d) 15.50 kg

134. A vacuum remediation process for cleaning soil uses air to remove the hydrocarbon. The soil contains 5 % by mass of hydrocarbon of average composition C_3H_7. The air removing the hydrocarbon is found to contain 3 % by volume of C_3H_7 after it has reached the analyzer tracking the dirty air. The kg of air needed to clean 1 kg of dirty soil is most nearly

(a) 1.08 kg
(b) 1.29 kg
(c) 28.13 kg
(d) 43.00 kg

135. The water that can be recovered from a groundwater stream is given by

(a) Flow rate of the stream.
(b) The water contained in the pores of the soil, i.e., porous fraction.

(c) The water contained in the pores that can be removed due to its unattended nature.

(d) Flow rate multiplied by the porosity of the soil.

136. Bioremediation is more useful for

(a) Heavier hydrocarbons
(b) Aliphatics
(c) Lighter hydrocarbons
(d) Chlorinated hydrocarbons

137. Water flows in the symmetrical trapezoidal channel lined with asphalt shown in the figure below. The channel bottom drops 0.1 m vertically for every 210 m of length.

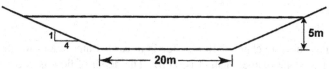

The water flow rate is most nearly

(a) 3 m^3/s
(b) 10 m^3/s
(c) 57 m^3/s
(d) 100 m^3/s

138. Water is to flow at a rate of 42 m^3/s in the concrete channel shown in the figure below.

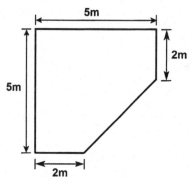

The vertical drop of the channel bottom per kilometer of length is most nearly

(a) 1
(b) 2
(c) 4
(d) 6

139. Water flows in the triangular steel channel shown in the figure below at a velocity of 3.8 m/s. The channel slope is 0.0025.

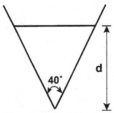

The depth of flow is equal to

   (a) 1.1 m
   (b) 3.8 m
   (c) 6.5 m
   (d) 10.1 m

140. A rectangular concrete channel 100 m side is to carry water at a flow rate of 110 m³/s. The slope of the channel is 0.001. The depth of flow is equal to
   (a) 0.01
   (b) 0.15
   (c) 0.25
   (d) 0.44

141. A corrugated metal pipe of 450 mm diameter flows half full at a slope of 0.003.

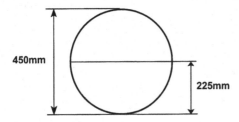

The flow rate is equal to

   (a) 0.13 m³/s
   (b) 0.17 m³/s
   (c) 0.25 m³/s
   (d) 0.45 m³/s

142. A 30-inch-diameter cast iron pipe on a 1/200 slope carries water at a depth of 4.0″ as shown in the figure below.

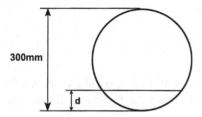

The flow rate (in ft³/s) is equal to

(a) 0.5
(b) 1.2
(c) 2.4
(d) 4.8

143. A 300-mm-diameter concrete pipe on a 1/400 slope is to carry water at a flow rate of 0.02 m³/s. The depth of flow in mm is equal to

(a) 100
(b) 200
(c) 325
(d) 405

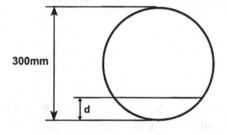
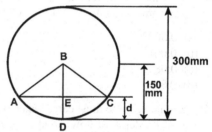

144. A 30-inch-diameter cast iron pipe on a 1/200 slope carries water at a depth of 4.0 in as show below.

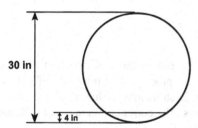

The flow rate utilizing figure (hydraulic elements of a circular section) is equal to

(a) 1.06 ft
(b) 1.56 ft
(c) 2.06 ft
(d) 2.50 ft

145. A 300-mm-diameter concrete pipe on a 1/400 slope is to carry water at a flow rate of 0.02 m^2/s, The depth of flow (utilizing the hydraulic elements of a circular section figure) is equal to

(a) 66 mm
(b) 100 mm
(c) 200 mm
(d) 300 m

146. Water is flowing in a concrete pipe with a velocity of 0.5 m/s at a depth of flow of one-tenth its diameter. The slope of the pipe is 0.005. The diameter is most nearly

(a) 0.04 m
(b) 0.64 m
(c) 0.84 m
(d) 1.00 m

147. A 20-inch-diameter concrete storm sewer pipe must carry a flow rate of 5 ft^3/s at a min velocity of 2.0 ft/s. The slope of the pipe is thus most nearly

(a) 0.0008
(b) 0.008
(c) 0.08
(d) 0.8

148. A concrete pipe must carry water at a slope of 0.005 at a velocity of 0.50 m/s and at a depth of flow equal to one-tenth its diameter. The pipe diameter is equal to

(a) 0.04 m
(b) 0.64 m
(c) 0.84 m
(d) 1.00 m

For problems 149 and 150, it is desired to reduce the NaCl content of water from 1.20 to 0.18 g/L using an electrodialysis process. The flow rate of the salt water is 20 m^3/min with a 48 % conversion. Each membrane has a surface area of 0.75 m^2 and each stack contains 100 cell pairs. The current efficiency is 84 %, and the current density is 4.2 mA/cm^2.

149. The membrane area required is equal to

(a) 1,000 m^2
(b) 3,000 m^2

(c) $5,000 \text{ m}^2$

(d) $10,500 \text{ m}^2$

150. The electrical power is equal to

(a) 36 kW

(b) 100 kW

(c) 500 kW

(d) 1,000 kW

151. A sample of 100 mL of waste liquid is poured into a crucible. The crucible is heated to 103 °C and weighted to be 45.454 g. The weight before heating is 46.118 g. Another sample of waste water (100 mL) is filtered and heated to 103 °C in a crucible. The weight of the crucible plus residue after heating to 103 °C is 46.194 g. The weight of this crucible plus liquid sample before heating is 46.247 g. The total dissolved solids in the sample are most nearly

(a) 53 mg/L

(b) 530 mg/L

(c) 1,530 mg/L

(d) 5,300 mg/L

152. For the above problem, the total suspended solids are equal to

(a) 664 mg/L

(b) 1,664 mg/L

(c) 6,100 mg/L

(d) 6,640 mg/L

153. The groundwater for a town is contaminated with a compound that is suspected to cause childhood leukemia. Nationally, leukemia strikes one in 10,000 children each year. In the town with the contaminated groundwater, 5 new cases of childhood leukemia are diagnosed in 2,000 children over a period of 5 years. The risk that is potentially attributable to the groundwater contaminant is

(a) 4 new cases in 10,000 children per year

(b) 5 new cases in 10,000 children per year

(c) 6 new cases in 10,000 children per year

(d) 24 new cases in 10,000 children per year

154. A process heater burns low-nitrogen fuel. The overall equation for the formation of thermal NO_x is

$$N_2 + O_2 + \text{heat} \rightarrow 2NO$$

At temperatures above 1,800 K, the rate of formation of NO is given by

$$[NO] = k_1 \exp\left(\frac{-k_2}{T}\right) [N_2][O_2]^{1/2} t$$

The brackets indicate mole fraction, k_1 and k_2 are constants, T is peak flame temperature in K, and t is the residence time of reactants at the peak temperature. If the concentration of oxygen is halved, formation of NO is

(a) decreased by 29 %
(b) decreased by 71 %
(c) increased by 41 %
(d) increased by 140 %

155. A temperature inversion is

(a) relatively cooler air trapped under warmer air
(b) relatively warmer air trapped under cooler air
(c) an effect of the Coriolis force
(d) caused by a stationary front

For problems 56 and 57, consider a chemical plant that uses a catalyst solution. This solution, when spent, is a RCRA-listed hazardous waste. The catalyst solution costs $0.30/kg. An alternative drop-in catalytic solution has been suggested that costs $2.09/kg. This catalyst becomes spent at half the rate of the original catalytic solution. The spent alternative solution can be managed as a non-hazardous waste at a cost of $0.22/kg.

156. Disregarding any additional monitoring, paperwork, and permit costs associated with the original solution, the cost for disposal of the original spent catalyst at which it becomes advantageous to switch to the non-hazardous alternative is

(a) >$0.44/kg
(b) >$0.86/kg
(c) >$2.01/kg
(d) >$2.85/kg

157. The additional monitoring, paperwork, and permit costs associated with disposing of the original solution as a hazardous waste are estimated to be one-third of the disposal fees. Taking this into account, the cost for disposal of the original spent catalyst at which it becomes advantageous to switch to the non-hazardous alternative is

(a) >$0.33/kg
(b) >$0.51/kg
(c) >$0.64/kg
(d) >$0.76/kg

158. Lime scrubbing is used to neutralize the sulfur dioxide in a stack. The chemical equations are

$$CaO\ (\text{lime}) + H_2O \rightarrow Ca(OH)_2$$
$$SO_2 + Ca(OH)_2 \rightarrow CaSO_3 + H_2O$$

The amount of sulfur dioxide that can be neutralized by 1 kg of lime is most nearly

   (a) 0.88 kg
   (b) 1.0 kg
   (c) 1.1 kg
   (d) 2.3 kg

159. A contaminant is released into groundwater, where its retardation factor is 3. The velocity of the groundwater is 10 m/d, and there is a well that is located in the direction of groundwater flow at a distance of 0.5 km from the contaminant source. The time at which the contaminant will reach the well is most nearly

   (a) 1.5 day
   (b) 15 days
   (c) 17 days
   (d) 150 days

160. An accident results in two railcars of hazardous waste breaking open and spilling onto a seep on the side of the tracks. The material in both railcars is of equal concern as a groundwater contaminant and as a human health and safety hazard. The retardation factor of the waste in railcar A is twice that of the retardation factor of the waste in railcar B. The waste in railcar A will travel through groundwater at the site at a rate that is most nearly

   (a) twice as fast as the waste in Railcar B
   (b) half as fast as the waste in Railcar B
   (c) the same as the waste in Railcar B
   (d) cannot be determined

161. The most difficult compound to incinerate in a stream is the one with the highest incinerability index. The equation for estimating the incinerability index is

$$I = C + \frac{100\,\frac{kcal}{g}}{H}$$

In this equation, C is mass percent and H is heating value. Compound concentration and heating values for a stream are

Compound	C, mass (%)	H, kcal/g
Acetonitrile	2.8	7.37
Benzene	30	10.03
Naphthalene	35	9.62
Vinyl chloride	1.0	4.45

The most difficult compound to incinerate in this stream is

(a) acetonitrile
(b) benzene
(c) naphthalene
(d) vinyl chloride

162. A cyclone's collection efficiency for particles of diameter 30 μm is required to be 90 %. The particle cut size for this cyclone is most nearly

(a) 3 μm
(b) 10 μm
(c) 30 μm
(d) 90 μm

163. A biohazard placard must be placed on the door to a room

(a) any time a biohazard agent is present
(b) only when biohazard agents in risk group 2, 3, or 4 are present
(c) only when biohazard agents in risk group 3 or 4 are present
(d) only when biohazard agents in risk group 4 are present

164. If the absolute pressure at the top of the water in a confined aquifer at sea level is 1.1 atm, the hydraulic head of the aquifer is most nearly

(a) 0.1 m
(b) 1.0 m
(c) 4.7 m
(d) 11 m

165. The hazardous waste treated in land treatment units is usually organic in nature because

(a) organic compounds are less toxic than other types of hazardous waste
(b) inorganic compounds are not as likely to be microbially or photolytically destroyed as organic compounds
(c) organic compounds are more likely to volatilize from the land treatment unit than inorganic compounds
(d) organic compounds are less likely to migrate to groundwater than inorganic compounds

166. The rotational speed of a fan doubles. All other parameters remain constant. The static pressure across the fan most nearly

(a) drops by one-half
(b) stays the same
(c) doubles
(d) quadruples

167. The indoor air pollutants that pose the greatest public health risk are

(a) formaldehyde and carbon monoxide
(b) radon and carbon monoxide
(c) radon and tobacco smoke
(d) tobacco smoke and formaldehyde

168. Federal agencies are required to assess the effects of proposed actions on historic properties under the provisions of the

(a) National Environmental Policy Act (NEPA)
(b) National Historic Preservation Act (NHPA)
(c) both NEPA and NHPA
(d) not applicable; federal agencies are not required to assess the effects of proposed actions on historic properties

169. A facility's releases have an objectionable smell. The ED_{50} for the facility's releases is 36 ppm. The nuisance threshold for the releases is 4 D/T. The concentration of the releases at the nuisance threshold is most nearly

(a) 7.2 ppm
(b) 9.0 ppm
(c) 40 ppm
(d) 140 ppm

170. The following is true of the meanings of precision and accuracy.

(a) The precision of an instrument reflects how nearly an instrument reading reflects the true value of what is being measured, while the accuracy of an instrument indicates the number of scientific digits in an instrument reading.
(b) The precision of an instrument indicates the number of significant digits in an instrument reading, while the accuracy of an instrument reflects how nearly an instrument reading reflects the true value of what is being measured.
(c) Both the precision and the accuracy of an instrument reflect the number of significant digits in an instrument reading.
(d) Both the precision and the accuracy of an instrument reflect how nearly an instrument reading reflects the true value of what is being measured.

171. Mixed wastes are wastes that

(a) are hazardous wastes regulated under the Resource Conservation and Recovery Act (RCRA) that are comprised of both solids and liquids
(b) contain radioactive wastes regulated under the Atomic Energy Act (AEA) and hazardous wastes regulated under the Resource Conservation and Recovery Act (RCRA)
(c) contain municipal solid waste and radioactive wastes regulated under the Atomic Energy Act (AEA)
(d) include both non-hazardous and hazardous solid wastes regulated under the Resource Conservation and Recovery Act (RCRA)

172. Aldrin, a pesticide, has a half-life in the environment of 2 years. Assume a first-order removal mechanism. It has been determined that a site where aldrin used to be manufactured cannot be developed into commercial property until the concentration of aldrin in the soil is 10 % of the current concentration. If no measures are taken to speed the removal of aldrin from the site, the length of time before development will be allowed at the property is most nearly

   (a) 1.4 years
   (b) 3.2 years
   (c) 6.6 years
   (d) 20 years

173. A method of measuring worker exposure to concentrations of benzene vapor is sought. The mean of the signals obtained when running blanks through one gas analyzer is 3.0 mA, with a standard deviation of 0.2 mA. Samples with certified reference concentrations of benzene result in the signals shown in the table. The concentration of benzene corresponding to the instrument's lower level of detection is most nearly

Concentration, ppm	Signal, mA
1	2.8
5	3.5
10	3.7
15	4.1
20	4.4

   (a) 5 ppm
   (b) 10 ppm
   (c) 15 ppm
   (d) 20 ppm

174. The collection efficiency of soot in an existing electrostatic precipitator is approximately 94 %. Collection efficiency must be increased to 98 %. To achieve the higher efficiency, the area of the electrostatic precipitator will be increased by most nearly

   (a) 4 %
   (b) 40 %
   (c) 76 %
   (d) 140 %

175. Two measured values, concentration and flow rate, are used to estimate the releases of hydrogen sulfide from a stack on a process that operates 350 days/year. The concentration in the stack has been measured to be 12 ppm with a standard uncertainty of 2 ppm and the flow rate has been measured to be 0.10 m^3/s with a standard uncertainty of 0.04 m^3/s. The standard uncertainty in the value for hydrogen sulfide released from the stack is most nearly

(a) 3.1 kg/year
(b) 3.7 kg/year
(c) 16 kg/year
(d) 24 kg/year

176. Residence time in minutes in a rotary kiln of length L (m), diameter D (m), kiln rake S (cm/m), and rotational speed N (rpm) is given by

$$t = \frac{19L}{DSN}$$

The residence time of a kiln must be doubled in order to meet new clean air regulations. This can be accomplished by

(a) halving the kiln rake and doubling the rotational speed while leaving the other parameters constant
(b) doubling the kiln rake while leaving the other parameters constant
(c) halving the kiln rake while leaving the other parameters constant
(d) doubling the diameter while leaving the other parameters constant

177. The density of air at sea level and at 20 °C is 1.2 kg/m^3, and oxygen comprises 21 % by weight of air. The change in the concentration of oxygen in the air at an altitude of 3 km above sea level compared to the concentration of oxygen at sea level is most nearly

(a) −30 %
(b) −10 %
(c) there is no change in the concentration of oxygen with elevation
(d) +40 %

178. The criteria air pollutants for which there are National Ambient Air Quality Standards are

(a) sulfur dioxide, carbon monoxide, benzene, and lead
(b) volatile organic compounds, particulate matter less than 10 μm in diameter, sulfur dioxide, nitrogen oxides, carbon monoxide, lead, and ozone
(c) sulfur dioxide, nitrogen oxides, carbon monoxide, ozone, and lead
(d) particulate matter less than 10 μm in diameter, sulfur dioxide, nitrogen oxides, carbon monoxide, ozone, and lead

179. The rate constant k can be expressed using the Arrhenius form

$$k = Ve^{-\frac{E}{RT}}$$

R is the universal gas constant, V is the frequency factor, T is temperature in Kelvin, and E is the activation energy. Assume a first-order reaction for destruction of a constituent in hazardous waste. At temperatures above 750 K, the frequency factor for the constituent is 28.5×10^6/s and its activation energy is 250 kJ/mol. The residence time needed to achieve 99.99 % destruction of the constituent in an incinerator whose temperature is kept at 1,600 K is most nearly

(a) 0.0005 s
(b) 1.7 s
(c) 47 s
(d) 2,800 s

180. Traffic through an intersection is expected to increase by 40,000 vehicles/day due to a proposed development. At this intersection, one would expect

(a) no change in the concentration of carbon monoxide
(b) a decrease in the concentration of carbon monoxide
(c) an increase in the concentration of carbon monoxide
(d) there is insufficient information

181. The unit risk for arsenic in drinking water is 5×10^{-5} $(\mu g/L)^{-1}$. Groundwater at a proposed development contains naturally occurring arsenic at a concentration of 1 ppb. Use of the groundwater as a drinking water supply will not be allowed if the estimated risk due to arsenic in the drinking water is greater than 1/10,000. It is determined that

(a) the groundwater may be used because at this concentration, the arsenic in the water poses a risk of less than 1/10,000
(b) the groundwater may not be used because at this concentration, the arsenic in the water poses a risk of greater than 1/10,000
(c) the number of residents in the development must be known in order to estimate the risk due to arsenic in the drinking water
(d) the drinking water may be used because the arsenic is naturally occurring

182. A sample containing methanol, n-hexane, 3-methylhexane, and 2-ethylhexane is sent through a gas chromatographer with a photoionization detector. There are four peaks in the resulting chromatograph. The peak with the longest retention time is much larger than the other peaks. The compound in this sample with the highest concentration is

(a) methanol
(b) 2-ethylhexane
(c) 3-methylhexane
(d) n-hexane

183. A worker is exposed to relatively high levels of acrolein in air for five minutes. He experiences discomfort in his eyes, which begin to water. This is an example of

   (a) acid deposition
   (b) a chronic health effect
   (c) a carcinogenic health effect
   (d) an acute health effect

   For problems 84 and 85, the sound pressure level at a point 1 m from a point source is 30 dB.

184. At this point, the sound pressure is most nearly

   (a) 9.0×10^{-5} Pa
   (b) 6.3×10^{-4} Pa
   (c) 0.0045 Pa
   (d) 0.02 Pa

185. The distance from the source at which this sound becomes inaudible is most nearly

   (a) 4.5 m
   (b) 28 m
   (c) 32 m
   (d) 1,000 m

186. If a trip blank tests positive for a compound of interest, it indicates that perhaps

   (a) the compound of interest is present in airborne form at the sampling site
   (b) the sampling container was contaminated with the compound
   (c) analytical equipment is not contaminated with the compound
   (d) the area where the sampling containers were prepared is not contaminated with the compound

187. Benzene is evaporating into a workroom at a rate of 0.01 g/s. The room is venting at a rate of 0.5 m^3/s and mixing in the room can be assumed to be ideal. The air in the room is at room temperature and the pressure is 1 atm. The concentration of benzene in the room is most nearly

   (a) 0.0063 ppm
   (b) 0.42 ppm
   (c) 6.3 ppm
   (d) 17 ppm

188. The four properties that determine if a waste is characteristically hazardous under the Resource Conservation and Recovery Act are

   (a) ignitability, persistence in the environment, reactivity, and carcinogenicity

   (b) corrosivity, reactivity, toxicity, and leachability of specific compounds at
      greater than threshold concentrations
   (c) ignitability, corrosivity, flammability, and toxicity
   (d) ignitability, corrosivity, reactivity, and leachability of specific compounds
      at greater than threshold concentrations

189. Air stripping is used to remove a pollutant from an aqueous stream. The air is
at 1 atm and 25 °C. The temperature change of air during the process can be
considered negligible. The concentration of pollutant in the water inlet to the
stripper is 80 ppm and the concentration of pollutant in the water leaving the
stripper is 4 ppm. The dimensionless Henry's Law constant for the pollutant
is 0.95. The average airflow rate is 0.1 m³/s, while the average water flow
rate is 0.02 m³/s. If the height of each transfer unit is 1 m, the height of the
column is most nearly

   (a) 4 m
   (b) 9 m
   (c) 10 m
   (d) 15 m

190. On an overcast night with a surface wind speed of 2 m/s, a compound is
emitted at a rate of 60 g/s from a stack whose effective stack height is 10 m.
The maximum concentration of the compound is found directly downwind of
the source and along the plume's centerline and is most nearly

   (a) 0.004 g/m^3
   (b) 0.005 g/m^3
   (c) 0.04 g/m^3
   (d) 0.05 g/m^3

191. The following measures result in a temporary reduction in the number of
mice and rats.

   (a) trapping and poisoning
   (b) rodent-proof trash containers
   (c) sanitation scheduling that prevents the overflow of trash
   (d) trimming foliage so that it does not overhang buildings

192. An individual standing 2 m from a source receives a dose of radiation. If this
person moves to a point 4 m from the source, the dose most nearly

   (a) drops to a fourth of the original dose
   (b) drops to a half of the original dose
   (c) stays the same
   (d) doubles

For problems 93 through 95, data points are 5.9, 6.4, 6.5, 6.9, 7.0, 7.2, 7.5, 7.9,
8.1, 8.5, 8.8, and 9.9.

193. The mean is most nearly

    (a) 7.2
    (b) 7.5
    (c) 7.6
    (d) 7.9

194. The standard deviation of these data points most nearly

    (a) 1.1
    (b) 1.2
    (c) 1.3
    (d) 1.4

195. The distribution of these points most nearly resembles a

    (a) log-normal distribution
    (b) normal distribution
    (c) uniform distribution
    (d) bimodal distribution

196. The concentration of methyl mercury in canned albacore tuna is 0.35 ppm. The ingestion reference dose for methyl mercury is 0.1 µg/kg/d. Assume canned albacore tuna is the only source of exposure to mercury. The number of 6-oz cans of tuna a 70 kg adult could consume in one year without exceeding levels assumed to be safe when developing the ingestion reference dose is most nearly

    (a) 1 can/year
    (b) 14 can/year
    (c) 43 can/year
    (d) 56 can/year

For problems 97 through 99, use a molecular weight for air of 29 g/mol. At any given moment, the flow of carbon dioxide into the atmosphere and the flow of carbon dioxide out of the atmosphere nearly equal each other.

197. The concentration of carbon dioxide in the atmosphere is approximately 360 ppm. The mass fraction of carbon dioxide in air is most nearly

    (a) 2.4×10^{-4} kg CO_2/kg air
    (b) 3.6×10^{-4} kg CO_2/kg air
    (c) 5.5×10^{-4} kg CO_2/kg air
    (d) 5.5×10^{2} kg CO_2/kg air

198. The total mass of air in the atmosphere is approximately 5.2×10^{18} kg. The total mass of carbon dioxide in the atmosphere is most nearly

    (a) 2.9×10^{15} kg CO_2
    (b) 9.5×10^{15} kg CO_2

(c) 2.9×10^{21} kg CO_2
(d) 1.1×10^{22} kg CO_2

199. Assume that the residence time of carbon dioxide in the atmosphere is 100 y. The flux of carbon dioxide into and out of the atmosphere is most nearly

(a) 3.4×10^{10} kg/day
(b) 5.1×10^{10} kg/day
(c) 7.9×10^{10} kg/day
(d) 2.9×10^{13} kg/day

200. The flow rate of benzene from a 50 m stack is 0.2 g/s. Wind speed, which in this case does not vary appreciably with altitude, is 6 m/s. The sky is overcast. The plume rise is 10 m. The ground-level centerline concentration 1 km downwind from the source is most nearly

(a) 0.00078 mg/m^3
(b) 0.0013 mg/m^3
(c) 0.0043 mg/m^3
(d) 0.026 mg/m^3

Exam 2 Solutions

101. The total solids are most nearly equal to the mass of residue left after heating the unfiltered wastewater to 103 °C, on a per liter basis. Now, 100 ml of wastewater upon heating to 103 °C (without being filtered) yield (46.480 − 46.048) g $= 0.432$ g $= 432$ mg.

Thus, 1 L of wastewater will yield $\frac{432 \text{ mg}}{100 \text{ ml}} \times \frac{1,000 \text{ ml}}{1 \text{ l}} = 4,320$ mg
Answer $= 4,320$ mg/L
The answer is (c).

102. The total suspended solids are equal to the (total solids (TS) minus the total dissolved solids (TDS)). The total solids (computed earlier) are equal to 4,320 mg/L. The total dissolved solids (TDS) computations are done using the data for the filtered water. The total dissolved solids (TDS) for 100 ml of filtered solution are equal to the weight of the residue remaining after heating to 103 °C

$$= 46.494 - 46.132 \text{ g}$$
$$= 0.362 \text{ g}$$
$$= 362 \text{ mg}$$

Therefore, on a liter basis, the TDS are equal to $\frac{362 \text{ mg}}{100 \text{ mL}} \times \frac{1000 \text{ mL}}{1 \text{ l}} = 3620$ mg/L.

Therefore, the total suspended solids (TSS)

$$= TS - TDS$$
$$= (4,320 - 3,620) \text{ mg/L} = 700 \text{ mg/L}$$

The answer is (b).

103. The total volatile solids are equal to the mg of residue remaining after heating to 600 °C on a per liter basis (of wastewater), for the case of unfiltered wastewater. This is equal to the mass of (dish + residue) minus mass of dish, for the case of unfiltered water (for 100 ml):

$$= 46.424 - 46.048 \text{ g}$$
$$= 0.376 \text{ g} = 376 \text{ mg}$$

Thus, for one liter of wastewater, total volatile solids
$$= \frac{376 \text{ mg}}{100 \text{ mL}} \times \frac{1000 \text{ mL}}{1 \text{ L}} = 3760 \text{ mg/L}$$
The answer is (d).

104. The volatile suspended solids (VSS) are equal to the (total volatile solids minus the volatile dissolved solids). The total volatile solids, computed earlier, are equal to 3,760 mg/L. The volatile dissolved solids for 100 mL of wastewater are obtained by computing the difference between the residue after 600 °C for filtered wastewater, which

$$= (46.556 - 46.132) \text{ g}$$
$$= 0.314 \text{ g} = 314 \text{ mg}$$

Thus, 1 L of wastewater contains $\frac{314 \text{ mg}}{100 \text{ mL}} \times \frac{1,000 \text{ ml}}{1 \text{ L}} = 3,140$ mg/L
Therefore, volatile suspended solids (VSS) = (3,760 - 3,140) mg/L
= 620 mg/L
The answer is (a).

105. The NEPA has a twin-pronged approach:

• Creation of the Council on Environmental Quality in the White House.
• Requiring federal scrutiny of all projects insofar as their effect on the environment.

This approach has led to NEPA having its reach into almost environmental issue in the United States. The other laws, while important, are more limited in their sphere of influence.
The answer is (d).

106. The growth of population is nearly linear between 1990 and 2000 and given by the formula:

$P = Po + At$

Taking 1990 as $t = 0$;
$P = $ Po in 1990
Thus, Po $= 100,000$
In year 2000 ($t = 10$, $P = 125,000$)
Therefore, $125,000 = 100,000 + At = 100,000 + (A)(10)$
Therefore, $A = \frac{125,000}{10} = \frac{25,000}{10} = 2,500$
Therefore, $P = (100,000) + (2,500)(t)$
Thus, in 2025, i.e., $t = 2025 - 1990 = 35$
$P = 100,000 + (2,000)(35) = 187,000$ persons

The per capital consumption of water grows at the rate of 1 % per year, compound basis. The compound interest formula needs to be used. Thus, per capita consumption in 2025 is equal to

$$(1.01)^{25}(40) = 1.2824 \times 40$$
$$= 57.30 \text{ gallons per person per day}$$

The total water consumption in 2025 is expected to be
$51.30 \times 187,500 = 9.5931$ million gallons per day
The answer is (d).

107. Using the compound interest formula, the expected population in 2020 is

$$= (1.02)^{20}(250,000)$$
$$= 371,487 \text{ persons}$$

The per capita consumption of water in year 2000 is

$$\frac{\text{total consumed}}{\text{population}} = \frac{10 \text{ million gallons per day}}{250,000 \text{ persons}} = \frac{10 \times 10^6}{2.5 \times 10^5} = \frac{40 \text{ gallons}}{\text{day-person}}$$

The per capita consumption of water in 2020 is

$$\frac{\text{total consumed in 2020}}{\text{population in 2020}} = \frac{16.7 \text{ million gallons per day}}{371,487 \text{ persons}} = 44.95 \frac{\text{gallons}}{\text{person-day}}$$

The compounded rate of increase per year of the per capita consumption can be found from
$(44.95) = (40)(x)^{20}$, where
$x = \left[(1)+\left(\frac{y}{100}\right)\right]$, where y % is the compound percent rate
Therefore, $x^{20} = \frac{44.95}{40} = 1.123$
Therefore, $\log x = \frac{1}{20}\log(1.123) = \frac{1}{20} \times 0.0506$
Therefore, $x = 10^{\frac{1}{20} \times 0.0506} = 1.0058$
Thus, compound percent increase $= (1.0058 - 1.0000) \times 100$ %

$$= 0.0058 \times 100 \text{ %}$$
$$= 0.58 \text{ %}$$

The answer is (b).

108. The volumetric flow rate

$$Q = \frac{C_v A_2}{\sqrt{1 - \left(\frac{A_2}{A_1}\right)^2}} \sqrt{\frac{2(p_1 - p_2)}{\rho}}$$

Now, $\dfrac{A_2}{A_1} = \left(\dfrac{D_2}{D_1}\right)^2 = \left(\dfrac{1}{2}\right)^2 = 0.25$

$\rho = 1 \text{ gm/cm}^3 = 1 \times 10^3 \text{ kg/m}^3$

$$A^2 = \frac{\pi D^2}{4} = \frac{(\pi)}{(4)} \left(\frac{1}{2}\right)^2 \text{ ft}^2$$

$$= 5.45 \times 10^{-3} \text{ ft}^2$$

$$= 5.45 \times 10^{-3} \text{ ft}^2 \times \left(\frac{30 \text{ cm}}{1 \text{ ft}}\right)^2 \times \frac{1 \text{ m}^2}{(100 \text{ cm})^2}$$

$$= 4.905 \times 10^{-4} \text{ m}^2$$

$C_v = 0.98$ for rounded venturi
$(p_1 - p_2) = 3$ mm of mercury
760 mm Hg = 1 atm pressure
$$= 1.01 \times 10^2 \text{ kPa}$$
$$= 1.01 \times 10^5 \text{ Pa}$$
$$= 1.01 \times 10^5 \text{ M/m}^2$$

Therefore, 3 mm Hg $= \frac{1.01 \times 10^5}{760} \times 3 = 398.68 \text{ N/m}^2$

Thus, $Q = \frac{\sqrt{0.98 \times 4.905 \times 10^{-4}}}{\sqrt{1 - 0.25}} \times \sqrt{\frac{2 \times 398.68}{10^3}} \text{ m}^3/\text{s}$

$$= \frac{4.015 \times 10^{-4}}{\sqrt{0.75}} \times 4.636 \times 10^{-4} \text{ m}^3/\text{s}$$
$$= 4.636 \times 10^{-4} \frac{\text{m}^3}{\text{s}} \times 3,600 \frac{\text{s}}{\text{h}}$$
$$= 1.66 \text{ m}^3/\text{h}$$

The answer is (a).

109. The velocity equation for the pitot tube is

$$V = \sqrt{\frac{2g(p_o - p_s)}{\gamma}}$$

where
V = velocity, m/s
p_o = stagnation pressure, N/m^2
p_s = static pressure, N/mw
γ = weight of fluid

Here, $(p_o - p_s) = (1.1 - 1)$ m of water
$= 0.1$ m of water

$$= \frac{0.1 \text{ m}}{0.34 \text{ ft} \times 0.3\frac{m}{ft}} \times 1 \text{ atm} \times 1.01 \times 10^5 \frac{N/m^2}{atm}$$
$$= 990 \text{ N/m}^2$$

$$\gamma = \rho g, \text{ where } \rho = 1 \times 10^3 \text{kg/m}^3$$

Therefore, velocity $V = \sqrt{\frac{2 \times -1.5pt3815pt0.3\text{pt}990}{1 \times 103 \times -1.5pt3815pt0.3\text{pt}}}$ ms
$$= 1.407 \text{ m/s}$$

The answer is (b).

110. Volumetric flow rate $= Q = CA\sqrt{2 \text{ gh}}$

Here, $h =$ height above the hole $= 10$ m
$A =$ area of hole

$$g = \frac{4}{\pi}(0.5)^2 \text{cm}^2$$
$$= \frac{4}{\pi} \times 0.25 \times 10^{-4} \text{ m}^2$$

$g = 9.8 \text{ m/s}^2$
$\quad C =$ coefficient of discharge
$\qquad = 0.98$, taken conservatively
Therefore, $Q = (0.98)\left(\frac{\pi}{4}\right) \times (0.25 \times 10^{-4})\sqrt{2 \times 9.8 \times 10}$
$= 2.70 \text{ m}^3/\text{s}$
$= 2.70 \times 60 \text{ m}^3/\text{min}$
Therefore, $Q = 161.5 \text{ m}^3/\text{min}$
The answer is (c).

111. The problem can be solved in tabular form.

Basis: 100 lb of solid waste

Component	Percentage of mass	Mass, lb_m	Percentage of moisture	Mass of water, lb
Plastics	22	22	0.5	11.00
Paper	42	42	22.1	9.28
Food waste	18	18	81.0	14.58
Cardboard	14	14	14.6	2.04
Other	4	4	25.2	1.00

$$\left(\begin{array}{l} \text{Computations in table:} \\ \text{Mass of water} = \dfrac{\text{Percentage of moisture} \times \text{Mass lb}_m}{100} \end{array} \right)$$

Thus, total mass of water in 100 lb of solid waste is equal to $(11 + 9.28 + 14.58 + 2.04 + 1.00) = 37.9$ lb
The answer is (c).

112. Tabular format.

Basis: 1 lb$_m$ of solid waste

Component	Percentage of mass	Mass $\frac{}{m}$	Density after compaction ρ	Volume = $\frac{m}{\rho}$
Plastics	22	0.22	6.2	0.035
Paper	42	0.42	22.1	0.019
Food waste	18	0.18	81.0	0.002
Cardboard	14	0.14	14.6	0.009
Other	4	0.04	25.2	0.0015

Space required = Total volume = $0.035 + 0.019 + 0.002 + 0.009 + 0.0015$
= 0.0665

113. Tabular format.

Basis: 100 lb of solid waste

Component	Percentage of Mass	Mass in (m) 100 lb of s.w.	Energy content E	mxE., Energy BTU
Plastics	22	22	14,200	31.24×10^4
Paper	42	42	9,100	38.22×10^4
Food waste	18	18	1,800	3.24×10^4
Cardboard	14	14	7,200	10.08×10^4
Other	4	4	7,600	3.04×10^4

Thus, the energy content of the solid waste is

$$= (31.24 + 38.22 + 3.24 + 10.08 + 3.04) \times 10^4 \text{ BTU/100 lb}$$
$$= 85.92 \times 10^4 \text{ BTU/100 lb}$$

The answer is (c).

114. The gas flux from the landfill is given by

$$N_A = \frac{-D\alpha^{4/3}(C_{atm} - C_{landfill})}{L}$$

$$N_A = \text{gas flux in} \frac{g.}{cm^2 s}$$

$$D = \text{Diffusivity of methane in the cover}$$
$$= 0.20 cm^2/s$$

$C_{landfill} = 0.001 \ g/cm^3$
Assume $C_{atm} = 0$, which happens when windy conditions prevail.
$L = $ Depth of cover material
$= 1 \ m = 100 \ cm$
Thus,

$$N_A = \frac{-(0.2)(0.21)^{4/3}(0.0 - 0.001)}{100}$$
$$= \frac{(0.2)(0.117)(0.001)}{100} g/cm^2\text{-s}$$
$$= 2.34 \times 10^{-7} g/cm^2 s$$

$\therefore$ for an area of 1 m^2 (= 10^4 cm^2) of cover materials, the amount of methane given out to the atmosphere

$$= 2.34 \times 10^{-7} \times 10^4$$
$$= 2.34 \times 10^{-3} g/s$$

The answer is (b).

115. The generation rates of methane and carbon dioxide have a ratio of 2:1. Assuming that the transfer rates outside the landfill are relatively small, the ratio of the concentrations of methane and carbon dioxide are 2:1 as well. Let concentration of methane be $2x$ and concentration of carbon dioxide be x. The rate of transfer of methane is

$$N_{CH_4} = \frac{-D\alpha^{4/3}(C_{A \ atm} - C_{A \ landfill})}{L}$$

$L = 1 \ m$
$\alpha = 0.18$
$C_{A \ atm} = 0$, assuming windy conditions
$C_{A \ landfill} = 2x$, assumed earlier
$D = 0.2 \ cm^2/s$

$$\therefore N_{CH_4} = \frac{-(0.2)(0.18)^{3/2}(-2x)}{1}$$
$$= +0.4(0.18)^{3/2}x$$

The rate of transfer of carbon dioxide is

$$N_B = \frac{-D\alpha^{3/2}(C_{B\ atm} - C_{B\ landfill})}{L}$$

$L = 1$ m
$\alpha = 0.18$
$C_{B\ atm} = 0$, assuming windy conditions
$C_{B\ landfill} = x$, assumed earlier
$D = 0.13$ cm^2/s

$$\therefore N_{CO_2} = \frac{-(0.13)(0.18)^{3/2}(-x)}{1}$$
$$= +0.13(0.18)^{3/2}x$$

$$\therefore \frac{N_{CH_4}}{N_{CO_2}} = \frac{(0.4)}{(0.13)} = \frac{40}{13} = \frac{4.0}{1.3}$$

The answer is (c).

116. The settled specific weight of the material in the lift is equal to (at $p = 2,000$ psi)

$$SWp = 1,100 + \frac{2,000}{0.014 + 0.0015 \times 2,000}$$
$$= 1,100 + \frac{2,000}{3.0014}$$
$$= 1,766.33 \text{ lb/yd}^3$$

Let A be the area of the bottom lift. The mass of the bottom lift is $(A)(5)(1100)\frac{\text{lb-m}}{\text{yd}^3}$

The volume of the lift after settlement is $\frac{(A)(5)(1100)}{1766.33}$ m $= 3.11$ m
Thus, the additional height available due to settling is $(5 - 3.11) = 1.89$ m
The answer is (b).

117. The stoichiometric equation can be written as

$$C_aH_bO_cN_d + \frac{(4a - b - 2c + 3d)}{4}H_2O \rightarrow$$

$$\rightarrow \frac{(4a + b - 2c - 3d)}{8}CH_4 + \frac{(4a - b + 2c + 3d)}{8}CO_2 + dNH_3$$

Here, $a = 50$, $b = 70$, $c = 30$, $d = 1$. The stoichiometric equation becomes

$$C_{50}H_{70}O_{30}N + \frac{(4 \times 50 - 70 - 2 \times 30 + 3 \times 1)}{4}H_2O$$

$$\rightarrow \frac{(4 \times 50 + 70 - 2 \times 30 - 3 \times 1)}{8}CH_4 + \frac{(4 \times 50 - 70 + 2 \times 30 + 3 \times 1)}{8}CO_2$$
$$+ NH_3$$

$$\therefore C_{50}H_{70}O_{30}N + \frac{(200 - 70 - 60 + 1)}{4}H_2O$$

$$\rightarrow \frac{(200 + 70 - 60 - 3)}{8}CH_4 + \frac{(200 - 70 + 60 + 3)}{8}CO_2 + NH_3$$

$$\therefore C_{50}H_{70}O_{30}N + 17.75H_2O \rightarrow 25.875CH_4 + 24.125CO_2 + NH_3$$

Molecular weight of the solid $= 12 \times 50 + 1 \times 70 + 16 \times 30 + 14$
$$= 600 + 70 + 480 + 14$$
$$= 1,164$$

Molecular weight of methane $= 12 + 4 = 16$.

From the stoichiometric formula, 1,164 kg of solid waste generates 16×25.875 kg of methane

$\therefore$ 1 kg of solid waste generates $\frac{16 \times 24.875}{1164}$ kg of methane, i.e., 0.355 kg of methane

The answer is (c).

118. (a), (b), and (c) are common sense definitions but not the definition of solid waste per RCRA.

The answer is (d).

119. The mass of solid waste that can be used for compost is $= 0.50 \times 500,000 = 250,000$.

The mass of sludge available $= 200,000$ kg/week. The N:C ratio of the resulting mixture is

$$= \frac{500.000\frac{kg}{week} \times 0.50 \times \frac{1}{15} + 200.000 \times \frac{1}{100}}{0.5 \times 500.000 + 200.000}$$
$$= \frac{18,666.67}{450,000} = 0.0414 \approx \frac{4}{100} \approx \frac{1}{25}$$

Thus, the C:N ratio is 25:1

The answer is (b).

120. The dry mass of solid waste which can be converted into compost

$= 0.5 \times 500{,}000 \times (1 - 0.45)$
$= 0.275 \times 500{,}000$
$= 137{,}000$ kg/week
The dry mass of sludge $= 200{,}000 \times 0.27 = 54{,}000$ kg/week
Thus, the total $= 54{,}000 + 137{,}500 = 191{,}500$ kg/week
The answer is (b).

121. Equivalent weight of $Ca^{2+} = 20.03$

Therefore, meq/L of $Ca^{2+} = \frac{80}{20.03} = 3.99$
Equivalent weight of $Mg^{2+} = 12.15$
Therefore, meq/L of $Mg^{2+} = \frac{160}{12.15} = 13.16$
Na^+ does not contribute to hardness; therefore, total hardness in meq/L is that of Ca^{2+} and Mg^{2+} combined.
$= 3.99 + 13.16 = 17.15$
The answer is (c).

122. Na^+ does not add to the total hardness. Since the pH is close to 7, the alkalinity is likely due to bicarbonate.

Component	Equivalent weight	Concentration		
	mg/meq	mg/L (given)	meq/L	mg/L as $CaCO_3$
Ca^{2+}	20.03	110	$\frac{100}{20.03} = 5.49$	$5.49 \times 50 = 274$
Mg^{2+}	12.15	100	$\frac{100}{12.15} = 8.23$	$8.23 \times 50 = 411.5$
Alkalinity	50	80	$\frac{80}{50} = 1.6$	$1.6 \times 50 = 80$

Total hardness is the sum of Ca^{2+} and Mg^{2+} (in terms of $CaCO_3$, as required by the question)

$$= 274 + 411.5$$
$$= 685.8 \text{ mg/L}$$

The answer is (c).

123. The non-carbonate hardness = (total hardness minus carbonate hardness). The carbonate hardness is equal to the alkalinity; unless the alkalinity is greater than the total hardness, in which case, the carbonate hardness is equal to the total hardness. In this case, the alkalinity = 80 mg/L as $CaCO_3$, which is less than the total hardness. Therefore, carbonate hardness = 80 mg/L as $CaCO_3$. Therefore, the non-carbonate hardness

= total hardness – carbonate hardness

Total hardness = 685.5 mg/L as $CaCO_3$. Therefore, non-carbonate hardness
= 685.5 − 80
= 605.5
The answer is (b).

124. Flux of water,

$$J_W = W_p(\Delta P - \Delta \pi)$$

Here, $W_p = 5.8 \times 10^{-7}$ g.mol/cm^2-s-atm
$\Delta P = P_{feed} - P_{perm}$
   = (50 − 2) atm
Therefore, $\Delta P = 48$ atmospheres
$\Delta \pi = 8$ atmospheres (given)
Therefore, $J_W = 5.8 \times 10^{-7}$ (48 − 8)

$$= 5.8 \times 10^{-7} \times 40 \text{ g.moles/cm}^2\text{-s}$$
$$= 2.32 \times 10^{-5} \text{ g.moles/cm}^2\text{-s}$$

Therefore, $J_W = 2.32 \times 10^{-5} \times$ (mol. wt of water) g/cm^2-s
$$= 2.32 \times 10^{-5} \times 18 \text{ g./cm}^2\text{-s}$$
Therefore, $J_W = 4.176 \times 10^{-4}$ g/cm^2-s
The answer is (c).

125.

$$J_S = K_p(C_{in} - C_{out})$$

Here, J_S = flux of salt

$$K_p = \text{permeance of NaCl} = 14.8 \times 10^{-6} \text{ cm/s}$$

Computation of C_{in} in g.mols/cm^{3}s
In 100 g of solution, there are 2.2 g of NaCl and (100 − 2.2) = 97.8 g of water.
Assuming the density of water is 1.0 g/cm^3, in 97.8 cm^3 of water, there are 2.2 g of NaCl.
Therefore, in 1 cm^3 of water, there are $\frac{2.2}{97.8}$ g of NaCl.
$= \frac{2.2}{97.8 \times 58.5}$ g.mols of NaCl (since molecular weight of NaCl = 58.5)
$= 3.85 \times 10^{-4}$ g.mols of NaCl
Computation of C_{out} of NaCl in g.mols/cm^3.
$C_{out} = 0.1$ g NaCl in (100 − 0.1) = 99.9 g of water.
Therefore, in 99.9 g of water, there are 0.1 g of NaCl.
Therefore, in 1 cm^3 of water, there are $\frac{0.1 \times 1}{18 \times 99.9}$ g.mols NaCl.
$= 5.56 \times 10^{-5}$ g.moles NaCl.
Therefore, $J_S = K_p (C_{in} - C_{out})$

$= 14.8 \times 10^{-6} \times (3.85 \times 10^{-4} - 5.56 \times 10^{-5} \times 58.5)$ g/cm^{3}s
Therefore, $J_S = 2.85 \times 10^{-7}$ g/cm^2-s
The answer is (a).

126. Now

$$u = \frac{-\partial x}{\partial y}$$

Therefore, $\chi = \int (-u)dy + C_1(x)$ where C_1 is a function of x.
Therefore,

$$\chi = - \int (4x + 6y)dy + C_1(x)$$
$$= -4xy + \frac{6y^2}{2} + C_1(x)$$

(1)

Also $v(x_1, y) = \frac{2\chi}{2x}$
Therefore, $\chi = \int v(x_1, y)dx + C_2(y)$ where $C_2 = $ function of y.
Therefore,

$$\chi = \int (-4y)dx + C_2(y)$$
$$= -4xy + C_2(y)$$

(2)

Equations (1) and (2) must be identical. Therefore, comparison of (1) and (2) yields

$$C_2(y) = 3y_2 + c$$
$$\text{and } C_1(x) = c$$

where $c = $ a numerical constant.
    Therefore, $\chi(x, y) = -4xy + 3y^2 + c$
    where c is a constant
    Since $\chi(0,0) = 0$
    Therefore, $0 = 0 + 0 + c$
    Therefore, $C = 0$
    Therefore, $\chi(x, y) = -4xy + 3y^2$
    The answer is (a).

127. The formula to be used is

$$Q = KA \left(\frac{dh}{dL} \right)$$

Since the head difference is 1.8 m and the distance between the wells is 1,200 m,

Therefore, $\frac{dh}{dL} = \frac{1.8}{1,200}$

The value of $K = 22$ m/d

The area A for the width of 1 m is equal to $(12 \text{ m})(1 \text{ m}) = 12 \text{ m}^2$

Therefore, $Q = 22 \text{ m/d} \times 1 \times 12 \text{ m}^2 \times \frac{1.8 \text{ m}}{1,200 \text{ m}}$ per m of width

Therefore, $Q = \frac{22 \times 12 \times 1.8}{1,200} \frac{\text{m}^3}{\text{day-m width}} = 0.396 \text{ m}^3/\text{day-m width}$

The answer is (b).

128. Pump-and-treat is used for cleanup of groundwater and not for soils.
The answer is (b)

129. $Q = 1,000 \text{ m}^3/\text{d}$

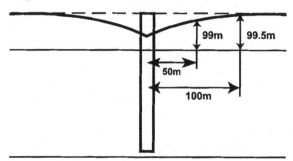

The formula for Q is

$$Q = \frac{2\pi \text{ KB } (h_1 - h_2)}{\ln\left(\frac{v_1}{v_2}\right)}$$

Here, $B = 50$ m, the thickness of the aquifer

$h_1 = 99.5$ m

$h_2 = 99$ m

$v_1 = 100$ m

$v_2 = 50$ m

Therefore, $Q \equiv \dfrac{(2\pi)(K)(50)(99.5 - 99)}{\ln\left(\dfrac{100}{50}\right)} = 1,000 \text{ m}^3/\text{d}$

Therefore, $K = \dfrac{(1,000)(\ln)\left(\dfrac{100}{50}\right)}{(2\pi)(50)(0.5)} = 4.4150 \text{ m/d}$

The answer is (c).

130. Formula used is

$$K = \frac{Q\ln\left(\frac{v_1}{v}\right)}{\pi\left(h_1^2 - h_2^2\right)}$$

The answer is (a)

Here, $Q = 1,000 \text{ m}^3/\text{day}$

$v_1 = 100$ m
$v = 50$ m
$h_1 = 49.75$ m
$h_2 = 49.25$ m

Therefore, $K = \dfrac{1,000 \times \ln\left(\dfrac{100}{50}\right)}{\pi\left[(49.75)^2 - [49.25]^2\right]} = 4.46$ m/d

The answer is (a).

131.

$$V = \frac{dh}{dx}$$

$K = 1.5 \times 10^{-2}$ m/min

$$\frac{dh}{dx} = 0.0015$$

Therefore, $v = 2.25 \times 10^{-5}$ m/s
$B = 40$ m

Therefore $\dfrac{Q}{2Bv} = \dfrac{0.1}{2(40 \text{ m})(2.25 \times 10^{-5})}$

$= 5.5$ m

Therefore, $\dfrac{Q}{Bv} = 2 \times 5.511.0$ m

Thickness $= \dfrac{Q}{2Bv}\left(1 - \dfrac{\varphi}{\pi}\right)$

Therefore $40 = 5.5\left(1 - \dfrac{\varphi}{\pi}\right)$

Therefore $\varphi = 2.3\pi$ radius

$$x = \frac{y}{\tan \varphi} = \frac{40}{\tan(2.3\pi)} = 29 \text{ m}$$

The answer is (d).

132. $Q = $ specific yield $\times$ volume of aquifer

$Q = 1,000$ m^3/day
Specific yield $= 0.2$
Therefore, volume of aquifer $= \dfrac{1,000}{0.2} = 5,000$ m^3/d
The answer is (d).

133. Basis: 1 kg of dirty soil. 1 kg of dirty soil contains 0.84 kg soil and 0.16 kg of
hydrocarbon. Assume that 98 % of hydrocarbon is removed.
   Therefore, amount of hydrocarbon removed $= 0.16 \times 0.98$ kg

$$= 0.156 \text{ kg}$$

The water–hydrocarbon solution contains 0.01 kg hydrocarbon, i.e., it contains
0.99 kg water and 0.01 kg hydrocarbon. Therefore, water needed to remove
0.156 kg of hydrocarbon

$$= \frac{0.99}{0.01} \times 0.156$$
$$= 15.4 \text{ kg water}$$

The answer is (d).

134. Basis: 1 kg mole of dirty air. 1 kg more of dirty air contains 0.97 kg moles air and 0.03 kg moles of hydrocarbon, since the ratio of volumes = ratio of moles at low pressure (1 atmospheric pressure). Average molecular weight of air = 29.00

Mass of air in 0.97 kg moles air = 29.0 × 0.97 kg
$$= 28.13 \text{ kg.}$$
Molecular weight of hydrocarbon (C_3H_7) = 3 × 12 + 7×1
$$= 36 + 7$$
$$= 43$$
Therefore, mass of hydrocarbon in 0.03 kg moles of hydrocarbon = 0.03 × 43 = 12.9 kg
Thus, total mass of 1 kg mole of dirty air = 1.29 + 18.13 = 29.42 kg
Thus, 1 kg mole of dirty air has removed 1.29 kg of hydrocarbon
Now 1 kg of dirty soil contains 0.05 kg hydrocarbon
Therefore, kg moles of dirty air resulting from removing 0.05 kg hydrocarbon

$$= \frac{0.05}{1.29} \times 1 = 0.0387 \text{ kg moles of dirty air}$$

Amount of air contained in 0.0387 kg moles of dirty air

$$= 0.0387 \times 28.13 \text{ kg} = 1.08 \text{ kg air}$$

The answer is (a).

135. **The answer is (c).**

136. Lighter hydrocarbons.

The answer is (c).

137.

$$v = \frac{1.0}{n} R^{2/3} S^{1/2}$$

$n = 0.015$
Hydraulic radius $R = \frac{A}{P_W}$

$$A = (20 \text{ m})(5 \text{ m}) + 2\left[\frac{(5 \text{ m})(3 \times 5 \text{ m})}{2}\right] = 175 \text{ m}^2$$

$$P_W = 20 \text{ m} + 2\sqrt{(5 \text{ m})^2 + (3 \times 5 \text{ m})^2} = 51.62 \text{ m}$$

Therefore, $R = \frac{175 \text{ m}^2}{51.62 \text{ m}} = 3.39$ m

$$= \frac{0.1 \text{ m}}{210 \text{ m}} = 4.76 \times 10^{-4}$$

Therefore, $V = \left(\frac{1.0}{0.015}\right)(3.39 \text{ m})^{2/3}(4.76 \times 10^{-4})^{1/2} = 3.28$ m/s
Therefore, $Q = (1.75 \text{ m}^2)(3.28 \text{ m/s}) = 574 \text{ m}^3/\text{s}$
The answer is (c).

138. Now,

$$V = \frac{1.0}{n} R^{2/3} S^{1/2}$$

$V = \frac{Q}{A}$, where Q is the flow rate

$$A = (5 \text{ m})(2 \text{ m}) + \left(\frac{2 + 5 \text{ m}}{2}\right)(5 - 2 \text{ m}) = 20.5 \text{ m}^2$$

Therefore, $V = \frac{42 \text{ m}^3/\text{s}}{20.5 \text{ m}^2} = 2.05$ m/s
Here, $n = 0.013$
Hydraulic radius $R = \frac{A}{P_W}$
Perimeter $P_W = 5 + 2 \text{ m} + \sqrt{(5 - 2 \text{ m})^2 + (5 - 2 \text{ m})^2} + 2 \text{ m} = 13.24$ m
Therefore, $R = \frac{20.5 \text{ m}^2}{13.24 \text{ m}} = 1.55$ m
Therefore, $2.05 \frac{\text{m}}{\text{s}} = \left(\frac{1.0}{0.0013}\right)(1.55 \text{ m})^{2/3} S^{1/2}$
Therefore, $S = 3.96 \times 10^{-6}$ m/meter of length
The answer is (c).

139. Now,

$$V = \frac{1.0}{n} R^{2/3} S^{1/2}$$

Given $V = 3.8$ m/s
Here, $n = 0.014$

Radius $R = \frac{h}{P_w} = \dfrac{2\left[\dfrac{(d)\left(d \tan 25°\right)}{2}\right]}{\frac{50}{\cos 20°}} = \bullet 17d$

Slope $S = 0.0025$
Therefore, $3.8 \text{ m/s} = \left(\frac{1.0}{0.014}\right)(0.17d)^{2/3}(0.0025)^{1/2}$

Therefore, $(1.065)3/2 = \left[(0.17d)^{2/3}\right]^{3/2}$

Therefore, $0.17d = 1.098$
Therefore, $d = 6.46$ m
The answer is (c).

140. Now,

$$V = \frac{1.0}{n} R^{2/3} S^{1/2}$$

Therefore, $V = \dfrac{Q}{A} = \dfrac{110 \ \text{m}^3/\text{s}}{(100 \ \text{m}) \, d} = \dfrac{1.1}{d} = \dfrac{\text{m}^2}{\text{s}}$

Here, $n = 0.013$

Slope $S = 0.001$

Radius $R = \dfrac{A}{P_W} = \dfrac{(100 \ \text{m}) \, d}{100 \ \text{m} \ + 2d}$

Therefore, $\dfrac{1.1 \ \text{m}^2}{d \ \ \text{s}} = \left(\dfrac{1.0}{0.013} \right) \left[\dfrac{(100 \ \text{m})(d)}{100 \ \text{m} \ + 2d} \right]^{2/3} (0.00025)^{1/2}$

Therefore, $d = 0.439$ m

The answer is (d).

141. Now,

$$V = \frac{1.0}{n} R^{2/3} S^{1/2}$$

Here, $n = 0.024$

Radius, $R = \frac{A}{P_W}$

Therefore, $A = \frac{1}{2} \left[\frac{(\pi)(.450 \ \text{m})^2}{4} \right] = 0.1767 \ \text{m}^2$

$$P_W = \frac{1}{2} \left[(\pi \cdot 52 \ \text{m}) \right] = 70.64 \ \text{m}$$

Therefore, $R = \dfrac{0.1767 \ \text{m}^2}{0.7069 \ \text{m}} = 0.2499$ m

Therefore, $V = \left(\dfrac{1.0}{0.24} \right) (0.2499 \ \text{m})^{2/3} (0.003)^{1/2} = 0.9055$ m/s

Therefore, $Q = (0.1767 \ \text{m}^2)(0.7366 \ \text{m/s}) = 0.1302 \ \text{m}^3/\text{s}$

The answer is (a).

142. Here,

$$V = \frac{1.486}{n} R^{2/3} S^{1/2}$$

Also, $n = 0.012$

Radius, $R = \frac{A}{P_W}$

$15'' - 4'' = 11''$ and $\sqrt{(15'')^2 - (11'')^2} = 10.2''$

Therefore, $\dfrac{11 \ \text{in}}{15 \ \text{in}} = 42.83°$

Therefore, Area $= \left[\dfrac{(\pi)(30 \text{ in})^2}{4}\right]\left[\dfrac{(2)(42.83°)}{360°}\right] = 168.19 \text{ in}^2$

Therefore, Area $= \left[\dfrac{(11 \text{ in})(10.2 \text{ in})}{2}\right] = 56.1 \text{ in}^2$

Therefore, Area (total) $= 168.19 \text{ in}^2 - (2)(56.1 \text{ in}^2) = 55.99 \text{ in}^2$

Perimeter $P_W = (\pi)(30 \text{ in})[(2)(42.83°)360°] = 22.43 \text{ in}$

Radius $R = \dfrac{55.49 \text{ in}^2}{22.43 \text{ in}} = 2.5 \text{ in} \Rightarrow 0.268 \text{ ft}$

Slope $S = 1/200$

Therefore, $v = \left(\dfrac{1.486}{0.012}\right)(0.208 \text{ ft})^{2/3}\left(\dfrac{1}{200}\right)^{1/2} = 3.08 \text{ ft/s}$

Therefore, $Q = (55.99 \text{ in}^2)\left(\dfrac{1 \text{ ft}^2}{144 \text{ in}^2}\right)(3.08 \text{ ft/s}) = 1.2 \text{ ft}^3/s$

The answer is (b).

143. Now,

$$v = \dfrac{1.0}{n}R^{2/3}S^{1/2}$$

Since, $v = \dfrac{Q}{A} = \dfrac{0.02 \text{ m}^3/s}{A}$

and $n = 0.013$

Slope $S = 1/400$

Therefore, $\dfrac{0.02 \text{ m}^3/s}{A} = \left(\dfrac{1}{0.013}\right)R^{2/3}\left(\dfrac{1}{400}\right)^{1/2}$

Therefore, $\dfrac{0.02 \text{ m}^3/s}{A} = 3.846R^{2/3}$

$AR^{2/3} = 0.0052 \text{ m}^3/s$

$$R = \dfrac{A}{P_W}$$

Therefore, $\dfrac{A^{5/3}}{P_w^{2/3}} = 0.0052 \text{ m}^3/s$

Length BE $= 0.15 \text{ m} - d$

and AE $= CD = \sqrt{(0.15 \text{ m})^2-(0.15 \text{ m} - d)^2}$

Angle $\angle ABD = \cos^{-1}\dfrac{0.15 \text{ m} - d}{0.15 \text{ m}}$

Therefore, Area $= \dfrac{(\pi)(0.30 \text{ m})^2}{4}\left[\dfrac{2\cos^{-1}\left(\dfrac{0.15 \text{ m}^2/s}{0.15 \text{ m}}\right)}{360°}\right]$

$$= 0.0039 \text{ m}^2\cos^{-1}\left(\frac{0.15 \text{ m} - d}{0.15 \text{ m}}\right)$$

Therefore,

$$\text{Area}_{\text{AECDA}} = (0.00039 \text{ m}^2)\left(\cos^{-1}\frac{0.15 \text{ m}-d}{0.15 \text{ m}}\right) - 2\left[\frac{-15 \text{ m}-d\sqrt{(0.15 \text{ m})^2-(0.15 \text{ m}-d)^2}}{2}\right]$$

$$P_W = \text{ADC} = (\pi)(0.30 \text{ m})\frac{\left(\cos^{-1}\frac{0.15 \text{ m}-d}{0.15 \text{ m}}\right)}{360°}$$

$$= (0.0052 \text{ m})\left(\cos^{-1}\frac{0.15 \text{ m} - d}{0.15 \text{ m}}\right)$$

$$\frac{\left[(0.00039 \text{ m}^2)\left(\cos^{-1}\dfrac{0.15 \text{ m} - d}{0.15 \text{ m}}\right) - (0.15 \text{ m} - d)\sqrt{(0.15 \text{ m})^2-(0.15 \text{ m} - d)^2}\right]^{5/3}}{\left[(0.0052 \text{ m})\left(\cos^{-1}\dfrac{0.15 \text{ m} - d}{0.15 \text{ m}}\right)\right]^{2/3}}$$

$= 0.01163 \text{ m}^3/\text{s}$
Therefore, $d = 0.305$ m or 305 mm
The answer is (d).

144. D = 30 in

$S = 1/200 = 0.005$ ft/ft
$Q_I = 11/4$ ft^3/s
$n = 0.012$

$$\frac{(Q_f)_{n=0.012}}{11.4 \text{ ft}^3/\text{s}} = \frac{0.013}{0.012}$$

$$(Q_f)_{n=0.012} = 12.4 \text{ ft}^3/\text{s}$$

$$\frac{(V_f)_{n=0.012}}{3.6 \text{ ft/s}} = \frac{0.013}{0.012}$$

$$(V_f)_{n=0.012} = 3.9 \text{ ft/s}$$

Therefore, $\dfrac{d}{d_f} = \dfrac{4 \text{ in}}{30 \text{ in}} = 13 \%$

Therefore, $\dfrac{Q}{Q_f} = \dfrac{4}{30} = 13 \%$

$$\frac{Q}{Q_I} = 3 \% \quad \frac{V}{V_f} = 40 \%$$

$Q = (0.03)(12.4 \text{ ft}^3/\text{s}) = 0.372 \text{ ft}^3/\text{s}$

Therefore, $v = (0.401)(3.9$ ft/s$) = 1.56$ ft/s
The answer is (b).

145. $D = 300$ mm

$Q = 0.02$ m^3/s
$S = 1/400$
$Q_f = 0.169$ m^3/s

$$\frac{Q}{Q_f} = \frac{0.02 \text{ m}^2/\text{s}}{0.169 \text{ m}^3/\text{s}} = 11.83 \%$$

Therefore, $\frac{d}{d_f} = 22 \%$
Therefore, $d = (0.22)(300$ mm$) = 66$ mm
The answer is (a).

146. The velocity $v = 0.5$ m/s, with slope $s = 0.005$. For this pipe, $n = 0.013$. The formula to be used is

$$v = \frac{1.0}{n} R^{2/3} S^{1/2}$$

Substituting

$$0.5 = \frac{1.0}{0.013} R^{2/3}(0.005)^{1/2}$$

Therefore, $R^{2/3} = \dfrac{(0.5)(0.013)}{(0.005)^{1/2}} = 0.019$

Therefore, $R = (0.19)^{3/2} = 0.026$ m
From graph, $\frac{d}{d_f} = 10 \% = 0.10$; $\frac{R}{R_f} = 25 \% = 0.25$

Therefore, $R_f = \frac{R}{0.25} = \dfrac{0.0026 \text{ m}}{0.25} = 0.0105$ m

$$R = \frac{D}{4}$$

Therefore, $D = 4R = (4)(0.0105)$ m $= 0.042$ m
The answer is (a).

147. $Q = Av$

Also, 5 ft^3/s $= (A)(2.0$ ft/s$)$
Therefore, $A = 2.5$ ft^2

Therefore, $A_f = \dfrac{(\pi)(2.5 \text{ ft})^2}{4} = 4.909$ ft^2

Therefore, $\frac{A}{A_f} = \dfrac{2.5 \text{ ft}^2}{4.909 \text{ ft}^2} = 50.93 \%$

Therefore, $\frac{d}{d_f} = 52 \%$
Therefore, $d = (0.52)(20$ in$) = 10.4$ in

Now, $v = \frac{1.486}{n} R^{2/3} S^{1/2}$

Given $v = 2.0$ ft/s

$n = 0.013$

$$\frac{R}{R_f} = 98 \%$$

Therefore, $R_f = \frac{P}{4} = \frac{2.0 \text{ ft}}{4} = 0.5$ ft

Therefore, $R = (0.98)(0.5 \text{ ft}) = 0.49$ ft

$$2.0 \text{ ft/s} = \left(\frac{1.486}{0.013}\right)(0.49 \text{ ft})^{2/3} S^{1/2}$$

Therefore, $S = 0.00079$

The answer is (a).

148. Now, $v = \frac{1.0}{n} R^{2/3} S^{1/2}$

Given $v = 0.50$ m/s

$n = 0.013$

Therefore, $0.50 \text{ m/s} = \left(\frac{1.0}{0.013}\right) R^{2/3} (0.005)^{1/2}$

Therefore, $R^{2/3} = 0.019$

Therefore, $R = 0.0026$ m

Therefore, $\frac{d}{d_f} = 10 \%$

Therefore, $\frac{R}{R_f} = 25 \%$

$$R_f = \frac{0.0026 \text{ m}}{0.25} = 0.0105 \text{ m}$$

Therefore, $R = \frac{D}{4}$

Therefore, $D = (4)(0.0105 \text{ m}) = 0.042$ m

The answer is (a).

149. Dilute stream flow rate:volumetric flow rate $\times$ area

$$Q = \frac{20 \text{ m}^3}{\text{min}} \times \frac{1 \text{ min}}{60 \text{ s}} \times 0.75 \text{ m}^2$$
$$= 0.25 \text{ m}^3\text{s}$$

$$\Delta c = \frac{1.2 \text{ g/L} - 0.18 \text{ g/L}}{58.5 \text{ g/mol}} = 0.0175 \text{ mol/L} = 17.4 \text{ mmol/L}$$

Membrane

$$A = \frac{(3)(F)(\text{flow rate})(\Delta c)}{(\text{Percentage of conversion})(\text{current efficiency})}$$

$$A = \frac{(1)(96,520)(0.25)(17.4)}{(Q48)(0.84)}$$

$A = 10413 \text{ m}^2$
The answer is (d).

150. Each membrane has a surface area of 0.75 m²; each stack contains 100 cell pairs.

Therefore, total area = (100)(0.75) = 75 m²
The number of stacks

$$\# = \frac{A}{\text{total area}} = \frac{10,413}{75} = 139$$

$$I = \frac{(1)(96,500)(0.25)(17.4)}{(Q48)(0.84)} = 4,997 \text{ A}$$

A/per stack = $\frac{4,997}{139}$ = 36 A
Electric power required

$$P = (I)(1)(100)(\text{pairs})$$
$$= (36)(1)(100)(139)$$
$$= 4,99,700 \text{ W}$$

$P = 500 \text{ kW}$
The answer is (c).

151. The total dissolved solids on a mg/L basis are those that are remaining in the crucible after filtered liquid is heated at 103 °C. Thus, the total dissolved solids in 100 mL of liquid are

$$= (46.247 - 46.194) \text{ g} = 0.053 \text{ g or } 53 \text{ mg}.$$

Therefore, the total dissolved solids in the sample, per liter of sample = 53 mg/ 100 mL × 1,000 mL/1 L = 530 mg/L
The answer is (b).

152. The total suspended solids are equal to the total fixed solids minus the total dissolved solids. Total fixed solids equal to amount of residue remaining for unfiltered liquid when it is evaporated at 103 °C. Thus, total fixed solids for 100 mL of liquid equals

$$= (46.118 - 45.454) \text{ g} = 0.664 \text{ g or } 664 \text{ mg}.$$

∴The total fixed solids per liter equals 664 mg/100 mL × 1,000 mL/ 1 L = 6,640 mg/L

∴Total suspended solids = total fixed solids minus total volatile solids

$$= (6,640 - 530) \text{ mg/L} = 6,110 \text{ mg/L}$$

The answer is (c).

153. The risk of childhood leukemia in the town with the contaminated groundwater is

$$\text{rate} = \frac{5 \text{ cases}}{(5 \text{ year})(2,000 \text{ children})} = 0.0005/\text{year}$$

The attributable risk is this value, minus the rate of disease in the population without the risk factor.

$$\text{attributable risk} = \frac{0.0005}{\text{year}} - \frac{1}{10,000 \text{ year}} = 0.0004/\text{year}$$

The answer is (a). Option (b) is incorrect because it is obtained by neglecting to subtract the background rate of disease. Option (c) is incorrect because it is obtained by neglecting to divide by the number of years during which new cases of childhood leukemia were identified in the town.

154.

$$\text{ratio} = \frac{\left(k_1 \exp\left(\frac{-k_2}{T}\right)[N_2]\left[0.5\,O_{2\text{original}}\right]^{1/2} t\right)}{\left(k_1 \exp\left(\frac{-k_2}{T}\right)[N_2]\left[O_{2\text{original}}\right]^{1/2} t\right)}$$

$$= \frac{(0.5)^{1/2}}{(1)^{1/2}} = 0.71$$

This means that the formation of NO is decreased by $(1 - 0.71) \times 100\,\% = 29\,\%$ when the concentration of oxygen is halved.

The answer is (a). Option (b) is incorrect because it is the fraction of NO produced before the oxygen concentration was halved, not the percentage decrease. Option (c) is incorrect because it is how formation of NO would be affected if the oxygen concentration were doubled. Option (d) is incorrect because it is the fraction of NO produced if the oxygen concentration were doubled, not the percentage increase.

155. Typically, the air temperature gets colder as altitude increases. This is inverted when the temperature rises as altitude increases (when cooler air is trapped under warmer air). Temperature inversions are not an effect of the Coriolis force; their causes include the advance of a warm front or a cold front and cold air filling a valley at night.

The answer is (a).

156. Let the rate of spent catalyst generation be r. The cost of purchasing and disposing of the catalytic solution is

$$C = \left(P_{solution} + P_{disposal}\right)r$$

For the original catalyst, the cost is

$$C_{original} = \left(0.30\frac{\$}{kg} + P_{disposal,original}\right)r_{original}$$

For the alternative catalyst, the cost is

$$C_{alternative} = \left(2.09\frac{\$}{kg} + 0.22\frac{\$}{kg}\right)(0.5)r_{original}$$

$$= \left(1.16\frac{\$}{kg}\right)r_{original}$$

Setting these costs equal to each other and solving for the price of disposing of the hazardous waste yields

$$\left(0.30\frac{\$}{kg} + P_{disposal,original}\right)r_{original} = \left(1.16\frac{\$}{kg}\right)r_{original}$$

$$P_{disposal,original} = 1.16\frac{\$}{kg} - 0.30\frac{\$}{kg}$$

$$= \$0.86/kg$$

The answer is (b). Option (a) is incorrect because it is double the cost of non-hazardous waste disposal of the alternative and does not take into account the difference in use or price of the solution. Option (c) is incorrect because it does not take into account the difference in solution use.

157. Again, let the rate of spent catalyst generation be r. The cost of purchasing and disposing of the catalytic solution is

$$C = \left(P_{solution} + P_{disposal} + P_{paperwork}\right)r$$

For the original catalyst, the cost is

$$C_{original} = \left(0.30\frac{\$}{kg} + P_{disposal,original} + \left(\frac{1}{3}\right)P_{disposal,original}\right)r_{original}$$

$$= \left(0.30\frac{\$}{kg} + 1.33P_{disposal,original}\right)r_{original}$$

For the alternative catalyst, the cost is unchanged and is

$$C_{\text{alternative}} = \left(2.09\frac{\$}{\text{kg}} + 0.22\frac{\$}{\text{kg}}\right)(0.5)r_{\text{original}}$$

$$= \left(1.16\frac{\$}{\text{kg}}\right)r_{\text{original}}$$

Setting these costs equal to each other and solving for the price of disposing of the hazardous waste yields

$$\left(0.30\frac{\$}{\text{kg}} + 1.33P_{\text{disposal,original}}\right)r_{\text{original}} = \left(1.16\frac{\$}{\text{kg}}\right)r_{\text{original}}$$

$$P_{\text{disposal,original}} = \frac{1.16\frac{\$}{\text{kg}} - 0.30\frac{\$}{\text{kg}}}{1.33}$$

$$= \$0.64/\text{kg}$$

The answer is (c). Option (d) is incorrect because it adjusts the price of the solution of the hazardous catalyst by 1.33 instead of adjusting the price of disposal of that catalyst.

158. First, find the overall chemical equation.

$$CaO + SO_2 \rightarrow CaSO_3$$

One mol of lime is consumed for every mol of sulfur dioxide that is neutralized. The molecular weights for lime and sulfur dioxide are

$$MW_{CaO} = 1\left(40\frac{\text{g}}{\text{mol}}\right) + 1\left(16\frac{\text{g}}{\text{mol}}\right) = 56 \text{ g/mol}$$

$$MW_{SO_2} = 1\left(32\frac{\text{g}}{\text{mol}}\right) + 2\left(16\frac{\text{g}}{\text{mol}}\right) = 64 \text{ g/mol}$$

The amount of sulfur dioxide neutralized by each kilogram of lime is

$$\frac{64 \text{ kg CaO}}{56 \text{ kg SO}_2} = 1.1 \text{ kg SO}_2/\text{kg CaO}$$

The answer is (c). Option (a) is incorrect because it is the result for lime consumed in neutralizing 1 kg of sulfur dioxide. Option (b) is incorrect because it is arrived at by failing to take the differences in molecular weights for lime and sulfur dioxide into account.

159. The retardation factor is the ratio of groundwater velocity to the contaminant's velocity. Therefore,

$$v_c = \frac{v_x}{R}$$

Also, distance is

$$d = v_c t = \frac{v_x t}{R}$$

Solve for t.

$$t = \frac{Rd}{v_x} = \frac{(3)(0.5 \text{ km})\left(1{,}000 \frac{m}{km}\right)}{\left(10 \frac{m}{d}\right)} = 150 \text{ days}$$

The answer is (d). Option (a) is incorrect because it is arrived at by neglecting to convert km to m. Option (c) is incorrect because it is obtained by reversing the ratio of velocities in the definition of retardation factor.

160. Because a spring or seep is a wetland that forms where the water table intersects with the soil surface, groundwater is immediately exposed to the hazardous waste in both railcars. If v_x is the average linear groundwater velocity, the velocity of the hazardous waste through the groundwater is

$$v_{\text{waste}} = \frac{v_x}{R}$$

The ratio of the velocity of the waste in railcar A to the velocity of the waste in railcar B is

$$\frac{v_A}{v_B} = \frac{R_B}{R_A} = \frac{1}{2}$$

The answer is (b).

161. Incinerability is determined from

$$I = C + \frac{100 \frac{\text{kcal}}{\text{g}}}{H}$$

For acetonitrile,

$$I = 2.8 + \frac{100 \frac{\text{kcal}}{\text{g}}}{7.37 \frac{\text{kcal}}{\text{g}}} = 16$$

For benzene,

$$I = 30 + \frac{100 \frac{\text{kcal}}{\text{g}}}{10.03 \frac{\text{kcal}}{\text{g}}} = 40$$

For naphthalene,

$$I = 35 + \frac{100 \frac{\text{kcal}}{\text{g}}}{9.62 \frac{\text{kcal}}{\text{g}}} = 45$$

For vinyl chloride,

$$I = 1.0 + \frac{100 \frac{\text{kcal}}{\text{g}}}{4.45 \frac{\text{kcal}}{\text{g}}} = 23$$

The compounds, in order of difficulty to incinerate (increasing incinerability index), are acetonitrile, vinyl chloride, benzene, and naphthalene.

The answer is (c).

162. The equation for calculating cyclone collection efficiency from particle cut size is

$$\eta = \frac{1}{1 + \left(\frac{d_{pc}}{d_p}\right)^2}$$

Solving for the particle cut size yields

$$d_{pc} = d_p \sqrt{\frac{1}{\eta} - 1}$$

$$= 30 \ \mu\text{m} \sqrt{\frac{1}{0.9} - 1}$$

$$= 10 \ \mu\text{m}$$

The answer is (b). Option (a) is incorrect because it is arrived at by neglecting to take the square root of the term.

163. When biohazard agents of risk group 2 or higher are present, a biohazard placard should be placed on the door where it can be easily seen.

The answer is (b).

164. The pressure at the top of the water in the aquifer is the atmospheric pressure plus the pressure of the water.

$$p_{\text{total}} = p_{\text{atm}} + p_w$$

Solve for p_w.

$$p_w = p_{\text{total}} - p_{\text{atm}} = 1.1 \ \text{atm} - 1.0 \ \text{atm} = 0.1 \ \text{atm}$$

The pressure of the water is

$$p_w = \rho g h$$

Solve for h.

$$h = \frac{p_w}{\rho g} = \frac{(0.1 \text{ atm})\left(1.015 \times 10^5 \frac{\text{Pa}}{\text{atm}}\right)}{\left(1{,}000 \frac{\text{kg}}{\text{m}^3}\right)\left(9.8 \frac{\text{m}}{\text{s}^2}\right)}$$

$$= 1.0 \text{ m}$$

The answer is (b). Option a) is incorrect because it is the result for the pressure at the top of the aquifer, not the hydraulic head in meters. Option (d) is incorrect because it is obtained by neglecting to remove atmospheric pressure before calculating hydraulic head.

165. The reason most wastes treated in land treatment units is organic in nature is because organic wastes are more likely to be microbially and/or photolytically destroyed or immobilized in a land treatment unit than inorganic wastes are. A waste cannot be applied to a land treatment unit unless it can be shown that the waste will be rendered non-hazardous or less hazardous.

The answer is (b). Option (a) is incorrect because it is not a true statement. Option (c) is incorrect because volatilization of the waste from the land treatment unit is not desirable. Option (d) is incorrect because it is not a true statement.

166. The static pressure across a fan is related to its rotational speed as follows.

$$\frac{\Delta p_{s1}}{\Delta p_{s2}} = \left(\frac{n_1}{n_2}\right)^2$$

Solving for $\Delta p_{s2}/\Delta p_{s1}$ gives

$$\frac{\Delta p_{s2}}{\Delta p_{s1}} = \left(\frac{n_1}{n_2}\right)^{-2} = \left(\frac{n_1}{2n_1}\right)^{-2} = 4$$

The answer is (d).

167. Of indoor air pollutants, radon and tobacco smoke present the highest level of concern for public health.

The answer is (c).

168. The National Historic Preservation Act (NHPA) requires federal agencies to consider the effects of proposed actions on historic properties. The effects of a proposed action on historic buildings and archeological sites must be avoided or mitigated. The National Environmental Policy Act (NEPA) requires the federal government to use all practicable means and measures to avoid environmental degradation; preserve historic, cultural, and natural

resources; and "promote the widest range of beneficial uses of the environment without undesirable and unintentional consequences."

The answer is (c).

169. Four dilutions/threshold, or 4 D/T, is equivalent to one-fourth of the threshold concentration, also called the ED_{50}.

$$C = \frac{ED_{50}}{\text{number of dilutions}} \frac{36 \text{ ppm}}{4} = 9 \text{ ppm}$$

The answer is (b). Option a) is incorrect because it assumes that there each unit of release is diluted with four units of air. Option (d) is incorrect because it assumes a concentration of the releases by a factor of four.

170. The precision of an instrument reflects the number of significant digits in an instrument reading. The accuracy of an instrument reflects how close the instrument reading is to the true value of what is being measured.

The answer is (b).

171. Mixed waste is defined as a waste mixture that contains both radioactive materials subject to the AEA and a hazardous waste component regulated under RCRA.

The answer is (b).

172. The time at which the concentration drops to 10 % of the original concentration can be calculated using the formula

$$C = C_0 \exp\left(-\frac{0.693t}{t_{1/2}}\right)$$

Solve for t.

$$t = -\frac{t_{1/2}}{0.693} \ln\left(\frac{C}{C_0}\right) = -\frac{2 \text{ y}}{0.693} \ln\left(\frac{0.1}{1}\right) = 6.6 \text{ years}$$

The answer is (c). Option (a) is incorrect because it fails to take into account the declining rate of destruction that occurs as the concentration declines. Option (b) is incorrect because it is arrived at by solving for t incorrectly.

173. The instrument's lower level of detection is the concentration that produces a signal equal to

$$\text{lower level of detection} = 2(1.645)s + m$$

In this equation, s is the standard deviation of the signal for the blank analyses and m is the mean of the signal obtained for the blank analyses. For this example, the lower level of detection is lower level of detection $= 2(1.645)(0.2 \text{ mA}) + (3 \text{ mA}) = 3.7 \text{ mA}$

The certified reference benzene sample at 10 ppm results in this signal, and 10 ppm is the instrument's lower level of detection for benzene.
The answer is (b).

174. The Deutsch–Anderson equation for electrostatic precipitator efficiency is

$$\eta = 1 - \exp\left(-\frac{WA}{Q}\right)$$

Solve for A.

$$\ln(1 - \eta) = -\frac{WA}{Q}$$

$$A = -\frac{Q}{W}(\ln(1 - \eta))$$

The area of the original electrostatic precipitator is

$$A_1 = -\frac{Q}{W}(\ln(1 - 0.94))$$

$$= 2.8\frac{Q}{W}$$

The area of the higher efficiency electrostatic precipitator is

$$A_1 = -\frac{Q}{W}(\ln(1 - 0.98))$$

$$= 3.9\frac{Q}{W}$$

The flow rate through the electrostatic precipitator and the terminal settling velocity of the particles does not change. The ratio of the areas is

$$\frac{A_2}{A_1} = \frac{3.9}{2.8} = 1.4$$

This represents an increase in area of 40 %.
The answer is (b). Option (c) is incorrect; it is the ratio obtained by dividing the area of the original electrostatic precipitator by the area of the more efficient one. Option (d) is incorrect because the ratio of the areas is not the increase in area required.

175. The formula for calculating the flow rate is

$$\dot{m} = \frac{CQ(\text{MW})}{22.4\frac{\text{L}}{\text{mol}}}$$

The standard uncertainty in the flow rate is

$$
\begin{aligned}
w_R &= \sqrt{\left(w_1 \frac{\partial f}{\partial x_1}\right)^2 + \left(w_2 \frac{\partial f}{\partial x_2}\right)^2 + \cdots + \left(w_n \frac{\partial f}{\partial x_n}\right)^2} \\
&= \frac{MW}{22.4\frac{L}{mol}} \sqrt{\left(w_C \frac{\partial \dot{m}}{\partial C}\right)^2 + \left(w_Q \frac{\partial \dot{m}}{\partial Q}\right)^2} \\
&= \frac{MW}{22.4\frac{L}{mol}} \sqrt{(w_C Q)^2 + (w_Q C)^2} \\
&= \frac{\left(1\left(32\frac{g}{mol}\right) + 2\left(1\frac{g}{mol}\right)\right)\left(3,600\frac{s}{h}\right)\left(24\frac{h}{d}\right)\left(350\frac{d}{y}\right)\left(\frac{1}{1,000}\frac{kg}{g}\right)}{\left(22.4\frac{L}{mol}\right)\left(\frac{1}{1,000}\frac{m^3}{L}\right)} \\
&\quad \times \sqrt{\left(\left(\frac{2}{10^6}\right)\left(0.10\frac{m^3}{s}\right)\right)^2 + \left(\left(0.04\frac{m^3}{s}\right)\left(\frac{12}{10^6}\right)\right)^2} \\
&= 24 \text{ kg/y}
\end{aligned}
$$

The answer is (d). Option (a) is incorrect because it is arrived at by assuming the ppm is on a mass basis, then multiplying the standard uncertainties and converting units. Option (b) is incorrect because it is arrived at by multiplying the standard uncertainties and converting units.

176. Kiln rake, diameter, and rotational speed are inversely proportional to residence time. Halving any of these while leaving the other parameters constant doubles the residence time.

    The answer is (c).

177. First, find the sea-level concentration of oxygen, using the mass fraction of oxygen in air, y_{O2}, and the total concentration of air at sea level, $X_{0, \text{total}}$.

$$
\begin{aligned}
X_{0,O_2} &= y_{O_2} X_{0,\text{total}} \\
&= (0.21)\left(1.2\frac{kg}{m^3}\right) \\
&= 0.25 \text{ kg/m}^3
\end{aligned}
$$

Now, find the concentration of oxygen at 3 km above the sea level

$$
\begin{aligned}
X_z &= X_0 \exp\left(-\frac{z}{8.400 \text{ km}}\right) \\
&= \left(0.25\frac{kg}{m^3}\right)\exp\left(-\frac{3 \text{ km}}{8.4 \text{ km}}\right) \\
&= 0.18 \text{ kg/m}^3
\end{aligned}
$$

This is a decrease of

$$1 - \frac{0.18\frac{g}{m^3}}{0.25\frac{g}{m^3}} = 30\%$$

The answer is (a). Option (d) is incorrect because the negative sign was left out in the exponential term.

178. The criteria air pollutants are particulate matter less than 10 μm in diameter, sulfur dioxide, nitrogen oxides, carbon monoxide, ozone, and lead. Volatile organic compounds have an effect on ozone concentrations and are sometimes regulated under the Clean Air Act, but there is no NAAQS for volatile organic compounds.

The answer is (d).

179. For a first-order reaction

$$\frac{dC}{dt} = -kC$$

Integrate to get

$$\int_{C_0}^{C} \frac{1}{C} dC = -k \int_{0}^{t} dt$$

$$\ln\left(\frac{C}{C_0}\right) = -kt$$

Solve for t and substitute the Arrhenius equation for k.

$$t = -\frac{\ln\left(\frac{C}{C_0}\right)}{V \exp\left(-\frac{E}{RT}\right)}$$

$$= -\frac{\ln\left(\frac{1 - 0.9999}{1}\right)}{\left(\frac{28.5 \times 10^6}{s}\right) \exp\left(-\frac{\left(250\frac{kJ}{mol}\right)\left(1,000\frac{J}{kJ}\right)}{\left(8.314\frac{J}{mol \cdot K}\right)(1,600 \text{ K})}\right)}$$

$$= 47 \text{ s}$$

The answer is (c).

180. Combustion of fossil fuels results in carbon monoxide releases.

The answer is (c).

181. The drinking water unit risk multiplied by the concentration in water is the risk of cancer from drinking the water.

$$\text{risk} = C(\text{unit risk})$$
$$= (1 \text{ ppb}) \left(\frac{1 \frac{\mu g}{L}}{1 \text{ ppb}} \right) \left(5 \times 10^{-5} \left(\frac{\mu g}{L} \right)^{-1} \right)$$
$$= 5 \times 10^{-5} = 0.00005 < 1/10,000 = 0.0001$$

The answer is (a).

182. Compounds separate in the chromatographer in order of their size, with the smaller compounds eluting first. Molecular weights of compounds in this sample are

$$MW_{CH_3OH} = 1\left(12\frac{g}{mol}\right) + 4\left(1\frac{g}{mol}\right) + 1\left(16\frac{g}{mol}\right) = 32 \text{ g/mol}$$
$$MW_{n-hexane} = 6\left(12\frac{g}{mol}\right) + 14\left(1\frac{g}{mol}\right) = 86 \text{ g/mol}$$
$$MW_{3-methylhexane} = 7\left(12\frac{g}{mol}\right) + 16\left(1\frac{g}{mol}\right) = 100 \text{ g/mol}$$
$$MW_{2-ethylhexane} = 8\left(12\frac{g}{mol}\right) + 18\left(1\frac{g}{mol}\right) = 114 \text{ g/mol}$$

The compound with the longest retention time is 2-ethylhexane, and since it has the largest peak, it has the highest concentration in the sample.
The answer is (b).

183. Acid deposition occurs when nitrogen oxides and sulfur dioxide in the atmosphere combine with moisture to create an acid that falls to the earth, either in precipitation or that is dry deposited. A chronic health effect is one that happens after long-term exposure, and a carcinogenic health effect is one that causes cancer. This is an example of an acute health effect.

The answer is (d).

184. Sound pressure level is given by

$$SPL = 10 \log_{10} \left(\frac{P^2}{P_0^2} \right)$$

Solve for P.

$$P = P_0 \sqrt{10^{SPL/10}} = \left(2 \times 10^{-5} \text{ Pa}\right) \sqrt{10^{30/10}} = 6.3 \times 10^{-4} \text{ Pa}$$

The answer is (b). Option (a) is incorrect because it is arrived at by taking the exponent of the natural number rather than 10. Option d) is incorrect because it is arrived at by neglecting to take the square root.

185. The change in sound pressure level with distance from a point source is

$$\Delta SPL = 10 \log_{10} \left(\frac{r_1}{r_2}\right)^2$$

Solve for r_2.

$$r_2 = \frac{r_1}{\sqrt{10^{\frac{SPL_2 - SPL_1}{10}}}} = \frac{1 \text{ m}}{\sqrt{10^{\frac{0-30}{10}}}} = 32 \text{ m}$$

The answer is (c). Option (a) is incorrect because it is arrived at by taking the exponent of the natural number rather than 10. Option (b) is incorrect because it is arrived at by assuming that 1 dB is the definition of inaudible. Option (d) is incorrect because it is arrived at by failing to take the square root.

186. A contaminated trip blank could indicate a contaminated sampling container, contaminated analytical equipment, or contamination in the area where the sampling containers were prepared. Because trip blanks remain sealed at the sampling site, they do not provide any indication of whether a compound of interest is present in airborne form at the site.

The answer is (b).

187. The equation for estimating the concentration of a steadily evaporating liquid is

$$C = \frac{Q_{vapor} R T_{air}}{k Q_{vent} P (MW)}$$

$$= \frac{\left(0.01 \frac{g}{s}\right) \left(0.08206 \frac{L \cdot atm}{mol \cdot K}\right) (298 \text{ K}) \left(\frac{10^6}{million}\right)}{(1) \left(0.5 \frac{m^3}{s}\right) \left(1,000 \frac{L}{m^3}\right) (1 \text{ atm}) \left(6\left(12 \frac{g}{mol}\right) + 6\left(1 \frac{g}{mol}\right)\right)}$$

$$= 6.3 \text{ ppm}$$

The answer is (c). Option (a) is incorrect because it was obtained by neglecting to convert from L to m^3 and from volume fraction to ppm. Option (b) is incorrect because it was obtained by using °C instead of K. Option d) is incorrect because it was obtained by dividing by the molecular weight of air instead of the molecular weight of benzene.

188. Characteristic RCRA hazardous wastes are classified as hazardous based on ignitability, corrosivity, reactivity, and leachability of specific compounds at greater than threshold concentrations.

The answer is (d).

189. The scrubber height is the number of transfer units multiplied by the height of the units, or

$$Z = (HTU)(NTU)$$

The number of transfer units is given by

$$NTU = \frac{R}{R-1} \ln\left(\frac{\left(\frac{C_{in}}{C_{out}}\right)(R-1)+1}{R}\right)$$

Here,

$$R = \frac{H'Q_A}{Q_W}$$

$$= \frac{(0.95)\left(0.1\frac{m^3}{s}\right)}{0.02\frac{m^3}{s}}$$

$$= 4.8$$

Therefore,

$$NTU = \frac{4.8}{4.8-1} \ln\left(\frac{\left(\frac{80\ ppm}{4\ ppm}\right)(4.8-1)+1}{4.8}\right)$$

$$= 3.5$$

The number of transfer units must be an integer and is equal to four.

$$Z = (NTU)(HTU) = (4)(1\ m) = 4\ m$$

The answer is (a).

190. Surface wind speed is defined as wind speed at 10 m above ground level, which in this case is equal to effective stack height. The maximum downwind ground-level concentration can be found either from a chart or by using an equation with curve-fitting constants. The equation provides a value in units of m^{-2} and is

$$\left(\frac{C\mu}{Q}\right)_{max} = \exp\left(a + b\ln H + c(\ln H)^2 + d(\ln H)^3\right)$$

The atmospheric stability class, which must be found in order to choose the proper curve-fitting constants, is D for overcast conditions regardless of wind speed, day or night. The constants are $a = -2.5302$, $b = -1.5610$, $c = -0.0934$, and $d = 0$. Solving for C_{max} yields

$$C_{max} = \frac{Q}{\mu} \exp\left(a + b \ln H + c(\ln H)^2 + d(\ln H)^3\right)$$

$$= \left(\frac{60 \frac{g}{s}}{2 \frac{m}{s}}\right) \frac{\exp\left(-2.5302 - 1.5610(\ln(10 \text{ m})) - 0.0934(\ln(10 \text{ m}))^2 + 0(\ln(10 \text{ m}))^3\right)}{m^2}$$

$$= 0.040 \text{ g/m}^3$$

The answer is (c). Option (d) is incorrect because the third logarithmic term in the equation is not squared.

191. Poisoning and to a lesser extent trapping of rats and mice results in a temporary reduction in rat and mice populations.

   The answer is (a).

192. The radioactive dose from a source is proportional to the square of the distance from the dose.

$$\text{flux at point 2} = \text{flux at point 1} \times \left(\frac{r_1}{r_2}\right)^2 = \text{flux at point 1} \times \left(\frac{2}{4}\right)^2$$

$$= \text{flux at point 1} \times 0.25$$

   The answer is (a).

193. The mean is the sum of the values divided by the number of values, or

$$\text{mean} = \frac{5.9 + 6.4 + 6.5 + 6.9 + 7.0 + 7.2 + 7.5 + 7.9 + 8.1 + 8.5 + 8.8 + 9.9}{12}$$

$$= 7.6$$

   The answer is (c).

194. The standard deviation is the square root of the sum of the difference between each point and the mean, divided by the number of data points, or

$$\text{standard deviation} = \sqrt{\frac{\begin{matrix}(5.9 - 7.6)^2 + (6.4 - 7.6)^2 + (6.5 - 7.6)^2 + (6.9 - 7.6)^2 \\ + (7.0 - 7.6)^2 + (7.2 - 7.6)^2 + (7.5 - 7.6)^2 + (7.9 - 7.6)^2 \\ + (8.1 - 7.6)^2 + (8.5 - 7.6)^2 + (8.8 - 7.6)^2 + (9.9 - 7.6)^2\end{matrix}}{12}}$$

$$= 1.1$$

   The answer is (a).

195. The distribution of these points most nearly resembles a normal distribution. Most of the points cluster around the mean, and the lowest value is nearly the same distance from the mean as the highest value. Tabulating the values into ranges yields the following result.

Range	Number of data points
4–4.9	0
5–5.9	1
6–6.9	3
7–7.9	4
8–8.9	3
9–9.9	1
10–10.9	0

Inspection of this tabulated data shows that the frequency increases around the mean and decreases away from the mean. Graphing these data would result in a bell-shaped curve with the mean near the peak of the bell. Also, one can show that approximately two-thirds of the data points lie within one standard deviation from the mean, and approximately 95 % of the data points lie within 2 standard deviations from the mean.

The answer is (b).

196. A hazard index greater than one is considered unacceptable. In this case, the hazard index is equal to the hazard quotient for ingestion of canned albacore tuna, or

$$HQ_{ing} = \frac{C_{ing}I}{RfD \times BW}$$

Set this equal to one and solve for I.

$$I = \frac{HQ_{ing} \times RfD \times BW}{C_{ing}}$$

$$= \frac{(1)\left(0.1\frac{\mu g}{kg \cdot d}\right)\left(365\frac{d}{y}\right)(70 \text{ kg})\left(\frac{1 \text{ can}}{6 \text{ oz}}\right)\left(\frac{1}{10^6}\frac{g}{\mu g}\right)\left(0.035\frac{oz}{g}\right)}{0.35\frac{g}{10^6 \text{ g}}}$$

$$= 43 \text{ cans/year}$$

The answer is (c).

197. The mass fraction of carbon dioxide in the air can be found from the concentration in ppm and the molecular weight of air and carbon dioxide. The molecular weight of carbon dioxide is

$$MW_{CO_2} = 12\frac{g}{mol} + (2)\left(16\frac{g}{mol}\right)$$
$$= 44 \text{ g/mol}$$

The mass fraction of CO_2 in the air is

$$y_{CO_2} = \left(\frac{360 \text{ CO}_2 \text{ molecules}}{10^6 \text{ air molecules}}\right)\left(44\frac{g \text{ CO}_2}{mol \text{ CO}_2}\right)\left(\frac{1 \text{ mol air}}{29 \text{ g air}}\right)$$
$$= 5.5 \times 10^{-4} \text{ kg CO}_2/\text{kg air}$$

The answer is (c). Option (a) is incorrect because it is obtained by multiplying by the molecular weight of air and dividing by the molecular weight of carbon dioxide. Option (b) is incorrect because it is mole or volume fraction. Option (d) is incorrect because it is obtained by neglecting to divide by a million.

198. The total mass of carbon dioxide in the atmosphere is

$$M = y_{CO_2} m_{total,air}$$
$$= \left(5.5 \times 10^{-4} \frac{\text{kg CO}_2}{\text{kg air}}\right)(5.2 \times 10^{18} \text{ kg air})$$
$$= 2.9 \times 10^{15} \text{ kg CO}_2$$

The answer is (a). Option (d) is incorrect because it is obtained by dividing the total mass of air by the concentration of carbon dioxide in the atmosphere.

199. To find the flux,

$$F = \frac{M}{\tau}$$
$$= \left(\frac{2.9 \times 10^{15} \text{ kg CO}_2}{100 \text{ y}}\right)\left(\frac{1}{365}\frac{y}{d}\right)$$
$$= 7.9 \times 10^{10} \text{ kg/d}$$

The answer is (c). Option (d) is incorrect because it fails to convert from days to years.

200. First, find the effective stack height. It is

$$H = h + \Delta h$$
$$= 50 + 10 \text{ m} = 60 \text{ m}$$

The Gaussian equation for estimating the concentration of a constituent in a plume is

$$C = \frac{Q}{2\pi\mu\sigma_y\sigma_z} \exp\left(-\frac{1}{2}\frac{y^2}{\sigma_y^2}\right)$$

$$\times \left(\exp\left(-\frac{1}{2}\frac{(z-H)^2}{\sigma_z^2}\right) + \exp\left(-\frac{1}{2}\frac{(z+H)^2}{\sigma_z^2}\right)\right)$$

At ground level along the centerline, y and z are zero, and this equation simplifies to

$$C = \frac{Q}{\pi\mu\sigma_y\sigma_z} \exp\left(-\frac{H^2}{2\sigma_z^2}\right)$$

Atmospheric stability class D is assumed during overcast conditions regardless of wind speed, day or night. Standard deviations can be read from curves for the vertical and horizontal dispersion of plumes and are σ_z= 32 m and σ_y = 73 m at x = 1,000 m and atmospheric stability class D. Therefore,

$$C = \frac{\left(0.2\frac{g}{s}\right)\left(1,000\frac{mg}{g}\right)}{\pi\left(6\frac{m}{s}\right)(32 \text{ m})(73 \text{ m})} \exp\left(-\frac{(60 \text{ m})^2}{2(32 \text{ m})^2}\right)$$

$$= 0.00078 \text{ mg/m}^3$$

The answer is (a). Option (b) is incorrect because it uses stack height instead of effective stack height. Option (d) is incorrect because it neglects the negative sign in the exponential term.

Printed in the United States
By Bookmasters